I0014662

Wolfgang Schreiner

Security and Privacy Management in Service Oriented Architectures

Wolfgang Schreiner

Security and Privacy Management in Service Oriented Architectures

Development of a Web Services Architecture for Managing Sensitive Data

VDM Verlag Dr. Müller

Impressum/Imprint (nur für Deutschland/ only for Germany)
Bibliografische Information der Deutschen Nationalbibliothek: Die Deutsche Nationalbibliothek
verzeichnet diese Publikation in der Deutschen Nationalbibliografie; detaillierte bibliografische
Daten sind im Internet über http://dnb.d-nb.de abrufbar.
Alle in diesem Buch genannten Marken und Produktnamen unterliegen warenzeichen-, marken-
oder patentrechtlichem Schutz bzw. sind Warenzeichen oder eingetragene Warenzeichen der
jeweiligen Inhaber. Die Wiedergabe von Marken, Produktnamen, Gebrauchsnamen,
Handelsnamen, Warenbezeichnungen u.s.w. in diesem Werk berechtigt auch ohne besondere
Kennzeichnung nicht zu der Annahme, dass solche Namen im Sinne der Warenzeichen- und
Markenschutzgesetzgebung als frei zu betrachten wären und daher von jedermann benutzt
werden dürften.

Coverbild: www.purestockx.com

Verlag: VDM Verlag Dr. Müller Aktiengesellschaft & Co. KG
Dudweiler Landstr. 99, 66123 Saarbrücken, Deutschland
Telefon +49 681 9100-698, Telefax +49 681 9100-988, Email: info@vdm-verlag.de
Zugl.: Wien, TU, Diss., 2007

Herstellung in Deutschland:
Schaltungsdienst Lange o.H.G., Berlin
Books on Demand GmbH, Norderstedt
Reha GmbH, Saarbrücken
Amazon Distribution GmbH, Leipzig
ISBN: 978-3-639-03496-7

Imprint (only for USA, GB)
Bibliographic information published by the Deutsche Nationalbibliothek: The Deutsche
Nationalbibliothek lists this publication in the Deutsche Nationalbibliografie; detailed
bibliographic data are available in the Internet at http://dnb.d-nb.de.
Any brand names and product names mentioned in this book are subject to trademark, brand or
patent protection and are trademarks or registered trademarks of their respective holders. The use
of brand names, product names, common names, trade names, product descriptions etc. even
without a particular marking in this works is in no way to be construed to mean that such names
may be regarded as unrestricted in respect of trademark and brand protection legislation and
could thus be used by anyone.

Cover image: www.purestockx.com

Publisher:
VDM Verlag Dr. Müller Aktiengesellschaft & Co. KG
Dudweiler Landstr. 99, 66123 Saarbrücken, Germany
Phone +49 681 9100-698, Fax +49 681 9100-988, Email: info@vdm-publishing.com
Copyright © 2008 VDM Verlag Dr. Müller Aktiengesellschaft & Co. KG and licensors
All rights reserved. Saarbrücken 2008

Printed in the U.S.A.
Printed in the U.K. by (see last page)
ISBN: 978-3-639-03496-7

Preface

The following thesis deals with the analysis and design of a software architecture with the purpose of supporting distributed user and document management. Particular attention is dedicated to the outsourcing of sensitive semi-structured document data to untrustworthy database servers and related security concerns. Basic technologies in the areas of Service Oriented Architectures (SOA) incorporated as Web services, as well as cryptography and access control are analyzed and brought into context with each other. Current development in the field of Web service security, especially standardization efforts and relevant scientific work, serve as starting point for the development of central security components with focus on sensitive data encryption as well as authorization of user requests on untrustworthy database servers.

Application settings in which the deployment of these technologies would engage an interesting role are identified, whereas main selection criteria comprise the applicability of multi-user systems and a high demand for security mechanisms for sensitive data. Use case scenarios in the area of human resource management are of special interest.

Furthermore, the thesis dwells on the realization of the concepts by referring to a prototypical reference implementation. A detailed specification of the developed interfaces shall facilitate future integration of the security services with existing application systems. Additionally a programming interface is provided for easy adaptation of the system depending on the requirements of the actual application. For rapid software development based on the technologies provided, a graphical user interface has been made available, which may be integrated into existing development environments. Results of the thesis are demonstrated and tested by sample applications, which are the result of practical works and master theses that have been carried out at the Vienna University of Technology.

Contents

List of Figures

List of Tables

Listings

Part I

Preliminaries

Chapter 1

Introduction

1.1 Problem Statement

Service Oriented Architectures (SOA) have in many ways revolutionized the art of software engineering and usage. Software is composed of a set of different components each providing different functionality. Service orientation introduces the concept of loosely coupled software components, since functionalities are uniformly described by service interfaces hiding the actual implementation. Standardized message protocols allow the integration of heterogeneous components independent of hosting platform or programming language. SOA is a concept, which can be realized by a variety of existing technologies, such as Remote Procedure Calls (RPC), the Distributed Component Object Model (DCOM) [109], the Common Object Request Broker Architecture (CORBA) [124] or Web services. The perspectives of complete independence in networking issues, including messaging, transaction management, data persistence or security, has lead to new perceptions in how software can be developed, deployed and actually used. Due to the extension of the Internet and accompanying reduction of costs and available bandwidth allowing high speed data transfer rates, user applications need no longer be installed locally on a private computer, but can be accessed remotely on an Application Service Provider (ASP) by a Web browser. Applications, such as Web mailers or Web hosting providers already embodied the ASP principles years ago, however, overall acceptance has just taken place due to the concept of social online communities based on the Web 2.0 principle and concrete success stories, including MySpace, YouTube or Flickr. Large companies like Google and Microsoft, tend towards providing their software as online services, which guarantees maximum accessibility as long as the servers stay online. All that is needed is a Web browser, which is available even on many mobile devices today. The trend towards remote software hosting brings a lot of benefits to application end-users, since they need no longer take care of file management. Precisely because users in some way lose control of their data, they need to trust the remote service provider that their data is secured from unauthorized access and modifications. This raises many security concerns, since at the outset the service provider cannot be trusted with respect to maintaining the privacy of user data. The following thesis addresses these security issues by developing a secure application architecture, which embodies fine granular access control for encrypted semi-structured data. Furthermore a service oriented framework is presented, which facilitates secure application development and specifies interoperability and interaction mechanisms.

1.2 Scientific Objectives

The motivation for the thesis is founded in the scientific efforts made by the SemCrypt research project, which is discussed at the end of Chapter 2 in greater detail. The motivation for SemCrypt stems from the increasing need of outsourcing huge amounts of complex data on external database servers with the possibility of accessing the data store anytime anyhow using standard communication protocols. In order to not be forced to care about securing the data store with additional hardware of software mechanisms, confidential data is stored encrypted. The application logic, which is capable of encrypting and decrypting the sensitive content, is moved to some trusted computing environment, either a trusted ASP or to the data owner itself. Sensitive data has to be dealt with in almost any application scenario imaginable, such as confidential customer or company data, but often users are short of confidence in the service provider and hesitate or refuse to reveal their private data. Who guarantees that data is properly secured from unauthorized access or not forwarded to any unknown third party? Among others, SemCrypt aims at developing a mechanism for ensuring that sensitive data can only be accessed by subjects, which are entitled to. Regardless of whether we are dealing with some single-user file storage or a multi-user collaboration platform, security can only be assured by encrypting data and restoring its plain text representation on the trusted client side using the client's valid private key, which is only known at the client's domain. When dealing with structured or semi-structured data it might be uneconomic to secure all of the information available, since only parts of it might require protection from unauthorized access. Furthermore, an access request that only refers to a small portion of the complete data set would cause an unnecessarily high performance overhead, since the whole database content would have to be loaded into main memory and analyzed computationally. Semi-structured data is mostly stored using XML or related technologies. Nowadays, XML usage is wide spread and has gained a key role for exchanging and storing electronic information. Many efforts for standardizing security issues with special focus on access control, signature and encryption have been made. With XML Encryption, for instance, it is possible to ensure confidentiality of XML structured information. The increasing deployment of XML for Web services and remote data management, however, requires additional methods for secure and efficient querying and updating of encrypted XML document data. The approaches provided in this thesis support state-of-the-art security standards and could thus add some significant value to increase the overall acceptance of E-Business applications and E-Collaboration models. Primary target markets for such techniques could be ICT outsourcing markets, which deploy advanced XML technology, and at the same time have serious shortcomings regarding appropriate security mechanisms. Furthermore, the xSP model may reduce risk by providing the possibility to apply Service Level Agreements (SLA). Especially, small and medium-sized businesses may benefit from the latter aspect.

Trust and security issues have a high impact on the success of inter-organizational business processes and are a crucial starting-point for improving the general acceptance of ICT business models. Trust relationships are essential between cooperating companies and customers. Since trust and security issues arising from document storage in the Semantic Web are mainly unsolved, especially, integration technologies in the area of XML and Web service standards are central subject to considerations regarding deployment of IT support in Small and Medium Enterprises (SME) with respect to their specific trust, security and efficiency requirements. Among others, trust and security issues are - although not yet sufficiently solved - basic requirements for collaboration in E-Business environments.

Another strong issue of motivation is data outsourcing in the context of Virtual Organizations (VO). VOs, Virutal Enterprises (VE) and Smart Business Networks [167] are concepts for networks of legally independent companies cooperating to produce complex products or services and are a new and major trend in the cooperative business (B2B) scenario. These three concepts are almost identical, though VEs are a particular molding of VOs. A VE is a temporary alliance of enterprises that cooperate closely in a service production process without having legal or financial interrelations to share skills or core competencies or resources in order to better respond to business opportunities, and whose cooperation is supported by computer networks

[37]. Different business models could be classified as VE. However, any such business model should have benefits for the customer as well as for the participating partners. The motivation to collaborate is to offer better services products to customers that cannot be offered by a single company. If a single company cannot easily handle complex demands in an effective manner, such networks may become useful. A VE enables outsourcing of responsibilities and is thus advantageous for large companies, which may then concentrate on their key capabilities, but also for small and medium companies, because they may participate in the production of very complex services and products. VOs defy the conventional rule for operating an organization. They do so by accomplishing tasks traditionally meant for an organization much bigger, more resourceful and financially stable by the means of collaborative efforts. Although such requirements exist since a long time, today the localization of appropriate partners and the management of the cooperation may be supported by recent information technology more efficiently. Of course, partners in a VE may also be non-commercial organizations. The success of VEs depends on the flexible, ad hoc establishment of cooperation and sometimes in the mass customization of delivered services. Cooperation in VEs can be established ad hoc without great organizational overhead due to ICT benefits. Especially better coordination of services and execution is an advantage that may be achieved by VEs and sophisticated ICT. On one side, this flexibility is achieved by electronic communication and information systems. For example, EDIFACT (ISO 9735) [166] has established as considerable success for large and stable networks such as in the food industry. However, for short-term cooperation such a standard is insufficient and moreover, too expensive for small and medium companies. To achieve the required flexibility, the usage of semantic knowledge in communication and cooperation is demanded [181].

A drawback in current ad hoc networks is missing trust, which consequently locks potential business partners away from information valuable for future collaborations. In B2C contracts, a customer may not want to reveal private data or preferences, to avoid the data being accessed by someone who is not entitled to. In B2B environments, participating companies may wish to hide knowledge about internal processes and be careful in establishing cooperation with unknown partners. Financial security is one aspect of trust, which may also be influenced by quality of service and cultural aspects. Total Quality Management (TQM) is a general management discipline demanding that processes and responsibilities are defined in an enterprise. In many traditional companies ISO 9000 certifications have gained tremendous importance. Digital signatures, certificates and encryption of information can increase trust relationships between single and network enterprises if their application is designed properly. Customers and employees may receive unique digital signatures to determine their identity in business processes unambiguously. Furthermore, processes must transparently specify whose duty it is to perform a certain task and these responsibilities must be controlled electronically. However, there may be portions of a process that should be visible only to certain groups such as partners or employees. If a well-known trust center certifies an organization, the trust into this organization will be higher. Thus an independent trust center may certify that a company really exists, has a certain reputation, is financial stable, its quality management complies with standards and achieves other organizational attributes that seem to be important. Moreover, delivered services should be certified. This is of course more complex in context with customized services and products in a VE. A very important issue in that context is the notion of trust. Several issues have been addressed by many experts in the field to improve existing trust frameworks. [191] assume that different partners have different trust levels assigned. They show how trust can be composed from individual trust levels for complex composite services in a VO. In this case, trust is assigned to single services based on the trust given to the organization or person that provides the service. A labeling scheme for sensitive objects within a VO domain is introduced in order to facilitate trust management. A label contains information about the originality of an object. The introduced framework also allows re-labeling of objects to react to the dynamic nature of a VO. With this model, recommendations supplied by other subjects can be made to adjust the trustworthiness of an object.

Nevertheless, the progressive use of XML for Web services as well as the need of remote data and document management using XML databases [27] additionally calls for special methods that efficiently and securely query and update (encrypted) XML documents at third parties that might not be trusted. Such techniques can make an important contribution to an increased acceptance and adoption of electronic business models and electronic collaboration in general as they address the trust problem, make use of the latest security standards, and even mitigate privacy concerns, if used appropriately. Consequently, there is a vast potential for these techniques in the ICT outsourcing markets, as these make use of XML and have large trust and security needs. Moreover, the service provider model and its markets offer many promising applications as risk is reduced and service level agreements (SLAs) can be slimmed, which makes this model more interesting for SMEs that have specific trust and efficiency requirements.

Chapter 2

Related Work

This chapter provides a discussion of relevant approaches and related work in the area of security management with respect to the work presented in this thesis. Starting with an introduction into the fundamentals of security management, including issues, such as authentication, authorization, cryptography or key management, the thesis provides a brief review of semi-structured data and related technologies and a summary of XPath fundamentals with respect to required security mechanisms presented in this work. XML plays an elemental role in information technologies today and is essential for understanding the outline of Web services in context with security and SOA, also presented in this chapter. The final section is devoted to the SemCrypt project, which scientific objectives and achievements are crucially important in order to provide a full understanding of the technologies developed and discussed in the course of the thesis.

2.1 Security Fundamentals

2.1.1 Distributed Systems Security

Security can be roughly divided into two parts: secure communication, which can be ensured via encryption and authentication, and authorization, often referred to as access control [159]. Encryption is discussed in the next section. It is a mechanism to ensure confidentiality and message integrity and pervades nearly every issue in security management. Confidentiality protects information from unauthorized access, while integrity deals with the problem that alterations to information can only be made in an authorized way. Message integrity ensures that information has not been tampered during transmission from sender to receiver. Authentication is about verifying the identity of a communication partner. Identity can thus be proved by the possession of a certain credential, for instance a password or a cryptographic key. While for confidentiality it is sufficient to encrypt the message content, integrity involves a notion of authentication to ensure that the message has not been modified and cannot be modified in the future. Authentication and message integrity are dealt with interchangeably, since both issues cannot do without the other one. What use is it to a receiver if a message has been transmitted correctly, if it cannot be sure that is has been sent by the entity it expects? And on the other hand what use is it to a receiver if it is sure about the sender's identity but not about the message content?

In distributed systems, four types of security threats need to be taken into account:

- Interception happens when a party gains unauthorized access to data

- Interruption refers to the situation in which data becomes unavailable

- Modification involves unauthorized message tampering

- Fabrication generates data which would normally not exist in an unauthorized manner

Secure systems should be immune against all possible security threats. This is usually achieved by formulating security policies, defining which actions entities or subjects, i.e. users, machines, etc., are allowed to perform. Policies are enforced by security mechanisms:

- Encryption implements confidentiality and integrity

- Authentication refers to entity identification

- Authorization is often used interchangeably with access control

- Auditing traces which subject accesses what when in which way

Designing secure distributed systems is a very sensitive area. Design flaws may very rapidly make security considerations obsolete. Performance is often a crucial issue in designing security protocols. Unfortunately, assumed optimization measures may have affect a protocol's correctness leading to unforeseen vulnerabilities, such as reflection or man-in-the-middle attacks. Secure channels deal with how to secure communication between interacting parties by providing confidentiality, integrity, authentication and authorization. The first issue to think about is the focus of control, dealing with the problem on which application components protection mechanisms should be applied. Data can be directly protected by prohibiting operations, which may cause invalid data states and violate data integrity. Other protection mechanisms affect unauthorized invocation of operations or take measures by which only specific people may gain access to the application independent of the operations they wish to perform. The second issue aims at which layer, referring to the ISO/OSI model, security mechanisms should be placed. Security on the transport layer is up to the local networking hardware, while data link, network and transport layers may be protected by the local operating system kernel. Application layer security refers to transportation of security information in the actual protocol headers, such as HTTP which lets the corresponding application decide what to do with the request. Security can be enforced at very different levels depending on the type of system and type of data that needs to be secured. For instance, messages between communication partners may be secured by encrypting the message itself or leaving the message in plain and encrypting the communication channel. Both approaches provide some notion of security but at different levels regarding the ISO/OSI model. This allows combination of security, since message may be encrypted and then be sent over the secure channel. Certainly, this only works if both communication partners implement the same type of security mechanisms.

2.1.2 Cryptography

Cryptography deals with message confidentiality and integrity. Transforming a plain text message into cipher text message using a cryptographic key, ensures that, given a strong encryption algorithm and key, an unauthorized party cannot gain access to the message, unless it somehow gets access to the required key. Integrity is ensured, since message tampering is only possible if the key is known as well. Alteration of the cipher text will be recognized as intrusion attempt as soon as the intended receiver tries to decrypt the message, which will result in a failure. Cryptographic systems are categorized into symmetric and asymmetric cryptosystems. Symmetric systems are also often referred to as shared key or secret key systems and use exactly one key for encrypting and decrypting a message. The key must be shared among interacting parties and locked away from unauthorized access. Asymmetric or public key systems consist of a key pair, namely a public key for encryption and a private key for decryption, where both keys must be mathematically related. The public key may be known by anyone, while the private key should be kept secret and only the party receiving a message should be able to decrypt and read the encrypted message. The advantage of symmetric keys over asymmetric keys is faster encryption/decryption computation. The difficult thing is key exchange without anyone else getting noticed. In practice key exchange using slow asymmetric encryption combined with fast symmetric encryption/decryption processing is used very often, for instance combining the public key Rivest-Shamir-Adleman (RSA) algorithm with the Data Encryption Standard (DES) [133]. Often it is required to prove that a message has been sent by one specific entity and no one else. This is something symmetric encryption cannot achieve, but is left to asymmetric cryptography. With asymmetric ciphers confidentiality can be achieved by letting the sender encrypt a message using the receiver's public key. Identity works the other way round, by letting the sender encrypt a message with the private key and all recipients decrypt with the sender's public key. This is comparable to a unique signature and ensures that the message could have only come from one specific sender. Unfortunately, public key encryption is very slow and not applicable to large documents. This is where hash functions come into play.

Hash functions are one-way mathematical functions, which take a plain text as input and calculate a fixed length hash value as output, usually 20 bytes, which allows a total of 2^{160} different values. This raises the probability of creating the same hash value out of two different messages to a 1 to 2^{160} chance, making guessing computationally infeasible. Hence, strong hash functions provide uniqueness, i.e. they are collision resistant in mathematical terms. The Secure Hash Algorithm SHA-1 and Message Digests MD4 and MD5, for instance have been proven to meet this requirement. Hash functions are very fast and useful for error detection, since as soon a message is altered, the hash-value a.k.a. checksum will be different. It is possible to generate a fixed-length encoded equivalent of the original input message, called the message digest, which in turn is easy to encrypt using public key cryptography to establish identity. However, what still is missing is the certainty that the owner of the public key is really the person who actually encrypted and sent the message. This is subject to key management discussed later on.

Security in cryptographic systems must be accomplished by using preferably strong cryptographic keys, not by keeping the ciphering algorithm secret. Algorithms are easy to break, strong keys are not. A good example is provided by the Global System for Mobile Communications (GSM) [120] security architecture. GSM uses relatively weak authentication and communication ciphers using shared key cryptography. If the algorithm could be kept secret, which was the consideration, it will be hard to ever break the security system. Because things that are secret often unintentionally attract increased attention, soon successful attacks on A5/1 and A5/2 algorithms used in GSM phones were published. However, algorithms should be made public to overcome proprietary ciphering implementations, which is especially important in distributed environments. Furthermore, algorithm publication leads to an improvement of the algorithm strength and quality. The XML Encryption standard, which is also subject of this chapter, specifies the usage of particularly strong algorithms, which have been intensively revised, such as Triple DES (3DES) or the Advanced Encryption Standard (AES). 3DES, for example, is based on the 25 year old DES standard, which is still considered strong and is widely applied, since it also runs fast in digital computing hardware. Cryptanalysis is a special field of research, which deals primarily with analyzing security systems and finding potential security leaks attackers could exploit to overcome protection mechanisms and invade the system.

2.1.3 Key Management

Establishment of cryptographic keys among interacting parties is not a trivial issue. Key exchange and distribution must be performed via secured channels at any rate. Sometimes it even has to be performed off the line. Also mechanisms for key invalidation and establishing new ones, known as key revocation, have to be provided. Since symmetric encryption algorithms are usually hundreds or even thousands of times faster than asymmetric algorithms, interacting parties use a shared secret key for encrypting and decrypting their messages. Distribution of shared secret keys is usually done via asymmetric keys. It is safe to encrypt a key with a public key and transmit it over the Internet, since it can only be decrypted using a special private key, which is not publicly known. The same approaches are applicable when interacting with a Key Distribution Center (KDC). A widely adopted algorithm is the Diffie-Hellman key exchange algorithm. If two parties wish to establish shared keys, they have to agree on two large numbers, which can be publicly known. At the same time both generate a number, which they keep secret. The Diffie-Hellman algorithm is based on the mathematical principle of modulo operations [133].

One problem of secret key management is scalability. If a system has a total of N hosts each communicating with each other using separate secret keys, N different keys are involved in the system with each host storing N-1 secret keys and leading to an overall key number of N*(N-1)/2, which needs to be managed by the system. A solution is a centralized KDC holding exactly N keys. In this approach each host first contacts the KDC to obtain a secret key it can use when initiating communication. Famous examples for

secret key authentication are the Needham-Schroeder protocol named after
its inventors and the Kerberos system developed at the Massachusetts Institute of Technology (MIT). Needham and Schroeder proposed the usage of
unique random numerical values attached to each message, so-called nonces
(number used once), to prevent replay attacks, i.e. sending a message twice
to exploit vulnerabilities in the security protocol. With public key authentication protocols, no KDCs are needed, since public keys need not be stored
in a secure manner and each host only needs to keep its own private key.

Another problem here is the credibility of the communication partner. Each
interacting party has to be sure that the public key really belongs to the
party who is claiming to be the owner. This can be accomplished via public
key certificates (PKC), which are hosted by a trusted Certification Authority (CA) and which are signed by the authority's public key. A certificate is
stored together with a public key and an identifier of the entity with which
the public key is associated. If a client wants to verify that a certificate indeed
belongs to the identified entity, it uses the authority's public key to validate
the public key and the identifier of the certificate. This works, since the client
trusts the CA that its public key has not been modified. Authorities' public
keys may in turn be validated using other CAs, leading to a hierarchical trust
model. Certificates often have limited lifetime. Certificate Revocation Lists
(CRL) are published regularly by a CA and indicate whether a certificate
has been revoked or not. KDCs and CAs play a significant role in public key
and secret key distribution processes. However big drawbacks are, they must
be trusted and provide a high level of availability, otherwise secure communication channels can not be established and public keys can not be verified.
Replication may be one workaround to providing availability, which on the
other hand makes a server more vulnerable.

2.2   XML and Semi-Structured Data

2.2.1   XML Essentials

In recent years, the eXtensible Markup Language (XML) [33] has become
the de-facto standard for data exchange. XML is the successor of the Standard Generalized Markup Language (SGML) [51]. It is a meta-language for
describing information, rather than handling information in an application
specific manner. Transformation languages, such as the eXtensible Stylesheet
Language (XSL) [45] define how XML data needs to be processed in order to
be used in an application specific context. For example: HTML provides an

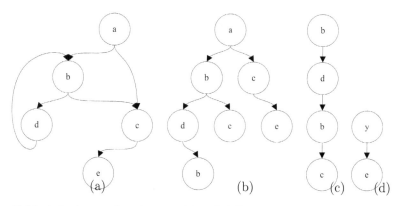

Table 2.1: A sample schema with valid XML document representations

element <ol/>, which tells a Web browser to display a consecutively num-
bered list of data, but does not provide any information about the type of
data. With XML one could describe an employee's professional qualifications as <Competencies/> pass it on to the XSL transformer, which in turn
generates a HTML document that could actually be rendered in the browser
window. XML thus abstracts data processing from its context-specific purpose, but is focused on semantic data annotations. Which elements are allowed in an XML document is defined by a separate source, called the XML
Schema [152] or a Document Type Definition (DTD) [173]. XML Schema is
a much more powerful language than DTD and allows very extensive document structure definitions, including names, types and contents of elements,
as well as cardinality and many more. The following illustration should serve
as clarifying though simplistic example: The left part of Table 2.1 (a) represents an exemplary XML Schema as cyclic, directed tree graph. It defines
elements a, b, c, d and e and their relations among each other. a is called
a global element, which directly refers to elements b and c. A schema may
contain cyclic references, enabling recursive occurrences of elements as in the
case of b and c. Unlike DTD, XML Schema does not prescribe a mandatory
root element, i.e. valid XML documents must not define one specific root
element to be schema valid. It may as well use only portions of a schema or
refer to multiple schemas, as schematically illustrated in graphs (b), (c) and
(d).

Exactly when referring to multiple schema instances, it might happen that
schemas define elements with the same name but different structures, re-

sulting in conflicts. The solution is namespaces. Namespaces refer to URIs
to distinguish overlapping element declarations from each other. They are
declared by the root elements of a schema and document to avoid parsing
errors during document processing. An XML Schema root or header looks
as follows:

Listing 2.1: Definition of an XML Schema file

```
<xsd:schema xmlns:xsd="http://www.w3.org/2001/XMLSchema"
    targetNamespace="http://semcrypt.ec3.at/xml/access/types"
    xmlns="http://semcrypt.ec3.at/xml/access/types"
    elementFormDefault="qualified">
    ...
</xsd:schema>
```

The root of all schemes is the `<schema/>` element. All XML Schema elements are per standard defined by the `http://www.w3.org/2001/XMLSchema`
namespace. The prefix `xmlns:xsd` declares `xsd` as namespace prefix to
be used for each element defined by the namespace URI, for instance the
`<schema/>` element. `xmlns` without suffix defines the default namespace. Elements declared therein must not be prefixed. The target namespace is often
abbreviated as `tns` and specifies the namespace for the elements defined by
the current schema. When setting `elementFormDefault` to qualified, all elements must be namespace qualified. A valid document header may then look
as follows:

Listing 2.2: A sample XML document

```
<XAccess xmlns="http://semcrypt.ec3.at/xml/access/types"
    xmlns:xsi="http://www.w3.org/2001/XMLSchema-instance"
    xsi:schemaLocation="http://semcrypt.ec3.at/xml/access/types
    xaccess.xsd">
    ...
</XAccess>
```

Namespace `xsi` is only required for the `schemaLocation` attribute, which
contains to parameters separated by white space. The first one indicates the
namespace to use and the other the physical location of the schema document,
in this case it resides in the same folder as the XML document. This is
important for validating the document against a schema. For integrating
multiple schemas, there are two possibilities, `import` and `include`:

Listing 2.3: Usage of import/include schema elements

```
<xsd:import (id)? (namespace)1 (schemaLocation)? ... />
<xsd:include (id)? (schemaLocation)1 ... />
```

A schema import only requires the definition of another namespace URI to be
added to the schema namespace. With include statements the `schemaLoc-
ation` is mandatory. Schemas to include must specify the same target names-
pace as the including schema.

2.2.2 XPath and XUpdate

XPath is for XML what SQL is for relational database tables. It can be used to select and navigate through portions of XML data. This section shall give an introduction to XPath features and XPath query processing, which is fundamental to describe access control for XML documents. The full specification of XPath can be found at [46] whereby the newer XPath 2.0 proposal [20] is part of the more complex XQuery standard [25]. XQuery can be used for much more than just querying XML, thus a comparison between XQuery and SQL would be maybe more appropriate. Related specifications are the XPointer [58] and XLink languages [59]. Referring to the previous section, an XML document is made up of different types of nodes, such as elements, attributes, text, comments and processing instructions. Depending on the corresponding schema, element nodes may be arbitrarily nested, while attributes can only be used to further specify an element. Text may occur within elements, attributes, comments or processing instructions. However, sensitive data is represented by elements and their contents, i.e. other elements, attributes or text. Comments may be used to further explain XML fragments and although it would be easily possible. Basic XPath syntax can be formally defined as:

XP := ("/"(N":")?E)* ("/"(N":")? "@"A)?

where E denotes an element contained by the affected XML document, A an attribute and N an optional namespace, the element or attribute respectively, are defined by. The syntax reminds of Unix file system navigation, where a single "/" selects the root directory and subsequent steps navigate down the directory hierarchy. A file, however represents a leaf and cannot contain other files or directories, which is also true for XML attributes. XPath queries can truly become very complex. The syntax contains very powerful features, which on the other hand are often redundant and may be rewritten by queries, which are equivalent but far simpler. One such mechanism is axes used to facilitate location based navigation. When an XML node, i.e. element or attribute, for instance, the node becomes the context node, which surrounding nodes may be more easily obtained using axis navigation. The default axis is child. Every path step that has no axis prefixed is assumed to be a child of the preceding step. The descendant axis contains all the descendants of the context node, except from attributes or namespace nodes, while descendant-or-self contains all the descendants of the context node, and the context node itself. Attributes and namespaces are excluded as well. attribute axes must be used in case the selected step

Axes	Abbreviated
child::N	N
descendant::N	//N
attribute::N	@N
self::N	.
descendant-or-self::N	descendant::N |.
namespace::N	N:
parent::N	..
ancestor::N	../../ etc.
ancestor-or-self::N	ancestor::N |.

Table 2.2: Axes and abbreviated equivalences used for XPath rewriting

points at an attribute node. The **self** axis contains just the context node, which is useful in cases the context nodes should be explicitly referenced. The **following-sibling** axis contains all the following siblings of the context node, and **following** points at all nodes that come after the context node in document order. Analogous, the **preceding** axis contains all nodes that occur before the context node in document order, excluding any ancestors of the context node, and again also excluding attribute nodes and namespace nodes. The **preceding-sibling** axis contains all the preceding siblings of the context node, unless context node is an attribute node or namespace node. The **namespace** axis references the namespace value of the context node if the node is an element. The **parent** axis contains only the parent of the context node, if there is one, while the **ancestor** axis contains all the ancestors of the context node, including its parents, grandparents, and so on. This axis always contains the root node unless the context node is the root node. The **ancestor-or-self** axis contains all the ancestors of the context node, and the context node itself.

Axes are a useful feature, but may be omitted using equivalent abbreviated syntax as shown in tables 2.2 and 2.3 [168]. The left column in Table 2.2 lists XPath axes and the right column their abbreviated equivalent. Note that preceding and following axes have been omitted, since they do not have an abbreviated equivalence and are more complex to handle. These issues are fundamentally important for query evaluation, which in turn forms the basis for the XML access control approach discussed in Chapter 5. It is crucially important to put these features into context with each other to map XPath expressions to a selected syntax portion and to be able to semantically compare XPath queries.

Examples		
/a/child::b	≡	/a/b
/a/descendant::b	≡	/a//b
/a/attribute::b	≡	/a/@b
/a/self::b	≡	/a/b/.
/a/descendant-or-self::b	≡	/a//b | /a/.
/a/b[namespace::N]	≡	/a/N:b
//a/parent::b	≡	//b/a
//a/ancestor::b	≡	//b//a
//a/ancestor-or-self::b	≡	//b//a | //a/.

Table 2.3: XPath rewriting examples

An XUpdate is defined as well-formed XML document, which makes extensive use of XPath [46] for specifying the XML portions to be updated. The following code sample inserts a new **Competency** element after the first **Competency** occurence, including a **name** attribute and a sub-element **CompetencyId** with attributes **id** and **idOwner**. The project Web site [162] provides additional example and usage descriptions to each update operation supported.

Listing 2.4: An XUpdate fragment creating an employee's competency

```
<xupdate:insert-after select="/SemCryptEmployeeInfo//Competency[1]" >
    <xupdate:element name="Competency">
        <xupdate:attribute name="name">
            Programming Language
        </xupdate:attribute>
        <xupdate:element name="CompetencyId">
            <xupdate:attribute name="id">
                C++
            </xupdate:attribute>
            <xupdate:attribute name="idOwner">
                9902261
            </xupdate:attribute>
        </xupdate:element>
    </xupdate:element>
</xupdate:insert-after>
```

XUpdate supports appending and inserting elements, attributes, text-content, comments and processing-instructions before and after the node-set selected by the specified XPath expression. Update, remove and rename operations are allowed as well. Furthermore, XUpdate introduces variables and includes conditional processing. The working draft was last updated in 2000, but its likely to be integrated into the XQuery specification in predictable future.

2.3   Web Services

The introductory chapter already established basic understanding of service orientation concepts. This section is focused on a special form of SOA, called Web services. Web services is a very generic and frequently discussed term referring to a technology, which is capable of implementing a wider range of distributed interaction mechanisms [34], where SOA is one of them. Often Web services are defined in the same context as SOA, though they do not necessarily imply each other. With distributed computing it has always been a major goal to achieve interoperability among interacting endpoints, regardless of programming languages the communication processes were written in or the operating system platform they were running on. Traditional middleware systems, like RPC and CORBA already achieved this, but required compatible communication partners, i.e. RPC and CORBA endpoints respectively. Web services evolved from existing distributed computing technologies and rely on a few core XML based specifications.

The UDDI consortium [123] defines Web services as

Definition 2.1. self-contained, modular business applications that have open, Internet-oriented, standards-based interfaces

As understood today Web services expose the functionalities of an information system and make it accessible through standard Web technologies. This reduces heterogeneity and facilitates application integration. Interoperability and application integration are very closely related. Application integration exists with many different characteristics most importantly for sharing and aggregating information inside and outside organization boundaries in the Business to Consumer (B2C) as well as in the Business to Business (B2B) sector [2].

The World Wide Web Consortium (W3C) [26] is very much involved in Web service specifications and definitions and considers Web services more precisely:

Definition 2.2. A Web service is a software application identified by a URI, whose interfaces and bindings are capable of being defined, described, and discovered as XML artifacts. A Web service supports direct interactions with other software agents using XML-based messages exchanged via Internet-based protocols.

or as

Definition 2.3. A Web service is a software system designed to support interoperable machine-to-machine interaction over a network. It has an interface described in a machine processable format (specifically WSDL). Other systems interact with the Web service in a manner prescribed by its description using SOAP messages, typically conveyed using HTTP with an XML serialization in conjunction with other Web-related standards.

The reason why conventional middleware failed to succeed was their complexity and incompatibility to other technologies. Roughly spoken, CORBA [124] requires compatible Object Request Brokers (ORBs) at each communication endpoint, DCOM [109] requires Microsoft Windows installed at each interacting host and RMI [155] requires Java processes for communications. Web services aim at overcoming these drawbacks by relying on XML standards for middleware functionalities, be it messaging, security or transaction support. The big advantage of XML is that it is a widely used meta-data format and that it is plain text rather than a binary format, which any computing system is able to handle. Web service standards are vastly independent from each other and from the specific underlying technology.

Conventional middleware suffers from the limitation of being centralized, i.e. controlled by a single company. B2B integration therefore requires companies to use the same middleware platform, including Workflow Management System (WfMS), Message Brokers for interaction or naming and directory servers. The same applies to security. Since companies rely on proprietary security mechanisms, required trust and confidentiality in company cooperations cannot be granted. Or considering transactions, traditional 2PC semantics as provided by most database management systems do not meet B2B requirements for long lasting transactions that may occur in Web service environments. Locking of resources for several days may be impractical in real-world cross-organizational distributed processes. Lack of standardization efforts made B2B integration costly. The EDIFACT standard was considered successful, but never widely adopted, because it was expensive to develop, difficult and almost impossible to maintain, such that only large companies could afford it. XML Web services aim at bypassing these shortcomings by defining middleware features at higher levels of abstraction, actually enabling independence from the underlying implementation. Web services may thus be easily integrated into more complex distributed applications, since compatibility among middleware platform is no longer an issue. Furthermore, Web services implement SOA following the "everything is a service" principle. Software should be composed out of loosely coupled, independent components. Many standardization efforts have been made with

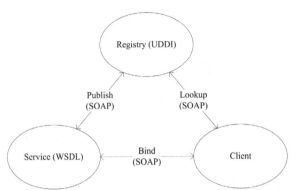

Figure 2.1: Common Web service infrastructure

the success of Internet technologies in mind. The World Wide Web works
due to standards without centralized coordination. Standardization is driven
by small, focused groups of companies and adopted by standardization organizations, such as OASIS or W3C, with the aftertaste of competing standards.

Technologies required for Web services comprise service description using
XML as common base language, service discovery for registering new services as well as searching for and locating services and service interaction
transparent of actual transport technologies. The traditional Web service
architecture [26] consists of a service provider describing a Web service in
a standardized way (WSDL) [44], a service registry to store service information (UDDI) [123] and a service consumer. Communication is performed
using SOAP [169], encoding protocols and interaction mechanisms into common XML syntax. The common SOAP data format enables loosely coupled
messaging or RPC, ignores semantics of complex communication patterns
and requires to be combined with an underlying middleware, such as HTTP
or SMTP. Describing is done via a common description language. The so-
called WSDL standard is comparable to the CORBA Interface Definition
Language (IDL), but separates interface description from other middleware
features, including addressing and protocol binding. Services provided via
WSDL may be invoked statically via precompiled stubs or dynamically using
stubs generated during runtime. UDDI implementations provide a naming
and directory interface for Web service environments. It allows publication,
location and pervasive usage of Web services at global scale.

Issues not handled by the standards described so far are subject to other proposals. A Web service specification only deals with very specific features, but does not attempt to tackle all of them as previous middleware platforms tried to achieve. WSDL, for instance, can only be applied for actual service description, but it does not provide information about invocation sequence of operations provided by the service described or describe conversational semantics and non-functional service properties. Latter ones may be contained by a service repository, such as UDDI. SOAP offers generic message templates, i.e. basic structure of message header and body, which may be enhanced by standards as WS-Security [122], adding security related information to SOAP. WS-Security by the way is one major subject of this thesis and will be discussed in greater detail shortly. Maintenance of conversation states or messaging sequences is transferred to meta-protocols, such as WS-Coordination. WS-Transaction [88] in turn extends coordination mechanisms with both short running and long living transactions. In the field of workflow management, the Business Process Execution Language (BPEL) [91] is very likely to succeed.

2.4 XML and Web Service Security

2.4.1 XML Encryption

The XML Encryption Working Group has published a proposal for encrypting and decrypting XML data and documents [89]. Its motivation is to exchange sensible data as encrypted XML data, in order to disable an illegal access to this data during communication. The proposal specifies the structure of encrypted data and the processing rules for decryption and encryption. An XML element is replaced by an encrypted cipher text, which again forms a well-formed XML element. A `type` attribute specifies whether the encrypted data is text, an XML element or the content of an XML element. Attribute encryption is not supported. The encryption algorithm is specified by the `EncryptionMethod` element and the key for the encryption algorithm can be specified with the `KeyInfo` element. The `CipherData` element contains either the encrypted text or a URI to the encrypted text. WS-Security uses XML Encryption [130] to hide selected information contained in SOAP messages. It additionally requires a reference list in the security header, which holds references to the encrypted data in the SOAP body. XML Encryption supports shared or secret key encryption, which requires both communication endpoints to know the key. An alternative approach is key wrapping, also called digital enveloping which uses shared key encryption to encrypt sensitive data and public key encryption for key exchange. Since the key is encrypted with the receiver's public key, the receiver is the only one who may decrypt the key and thus apply it on the SOAP message body. When using key wrapping, WS-Security additionally stores the wrapped key in the security header and the reference list is contained by the encrypted key data structure.

2.4.2   XML Signature

A number of activities were performed to consider security concepts in the context of the Semantic Web and XML. The XML Signature Working Group [65] has published 2002 a proposal for a digital signature for XML documents and elements [18]. The motivation for a digital signature is to proof the integrity of a document, i.e. nobody may alter unintentionally or deliberately the document, without detection of this alteration by the intended receiver and an unambiguous identification of the sender of a message. The adopted proposal describes the structure of a digital signature and a number of processing rules to be applied if a message is signed or if the integrity is checked. A signature may be part of a document, it may embrace the document or the signature may be outside of the document. A signature may also be used for more than one object. A signature may specify several processing steps performed before the signature is computed. These steps may be a canonicalization of an XML document, an XSLT-transformation and/or a filtering with XPath [31]. The proposal specifies some algorithms for a digital signature in its name space to be used. Further algorithms may be used if an URI is given where the specification of the algorithm is defined. An optional element can be used to supply key or certificate with the signature. Digital signatures in SOAP documents can be used for verifying a security token or the message integrity by comparing document hash values.

Message timestamps are the only new mechanism that WS-Security introduces to SOAP messages. All the other features are a way to incorporate existing security technologies into Web service messaging. If message timestamps are used, the sender may include a field, which tells the receiver about the expiry date of the message, which tells the receiver whether to process or discard the message. Furthermore it tells the processing intermediaries when the message was created and additionally allows timestamp based tracing.

We have now seen that WS-Security comprises a lot of issues that may be used with SOAP messaging. Of course, it allows maximum flexibility when configuring the security layer of a Web service, but always with the drawback of tremendous performance tradeoffs. Using XML Signature, XML Encryption and security tokens to which extent ever, may dramatically increase the size of a SOAP message. Not to mention other middleware features, such as coordination, transaction or context information, that might be necessary in an application setting, a SOAP header is very likely to quickly explode in size. This may appear a bit weird especially if the size of the body containing the actual message is confined to a few lines. In large inter-enterprise application software message size could significantly affect transmission time, processing time and memory consumption. In particular XML processing, which requires canonicalization, and computationally expensive signature validation or asymmetric encryption/decryption algorithms may significantly slow down the communication process.

2.4.3 The WS-Security Stack

WS-Security is not a new technology per se but rather incorporates existing security mechanisms within SOAP messages [96]. It is important to understand that WS-Security directly operates on the messages themselves, i.e. on the application layer. Security is very complex and may also be directly applied on the transport level, which leaves the message unsecured on any higher level in the ISO/OSI model [132]. For example, it is fine to use existing transport layer security mechanisms, such as Secure Sockets Layer (SSL), to Web service messages since they do not care which type of message they encrypt. Moreover they are simple, supported by most systems and well known to system administrators. However, there are a number of serious drawbacks to the approach: Transport layer security only applies to a certain transport protocol, for instance SSL cannot be applied to a protocol other than HTTP, messages cannot be accessed during transport (at least partially) but are in the clear after leaving the receiver's SSL socket, and we must apply the same security for all of the messages being carried over the secure pipe. Pros and cons of transport layer security are almost inverted in case of message layer security. WS-Security allows securing parts of each message differently, which makes it possible to encrypt the message content with a private key for the message receiver and using a separate key for the routing information. This is an issue also addressed by WS-Routing [110]. Incorporation with other emerging Web service standards and involving flexibility is the most promising feature of Web service security but at the same time its probably biggest weakness. Since Web services aim at creating an abstract middleware and transport is not part of any middleware specification, WS-Security thus addresses the issue of securing Web service messages themselves on the application layer while staying transparent to the underlying transport technology. Before becoming a standard proposed by IBM,

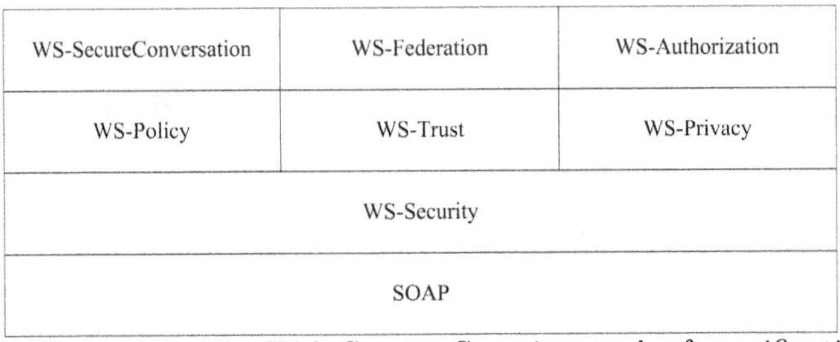

WS-SecureConversation	WS-Federation	WS-Authorization
WS-Policy	WS-Trust	WS-Privacy
WS-Security		
SOAP		

Figure 2.2: The Web Service Security stack of specifications

Microsoft and VeriSign in 2002, WS-Security was defined as SOAP extension
supporting only digital signatures. The following figure was taken from [118],
a whitepaper published by Microsoft and IBM: Similar to other Web service
standards, WS-Security defines the type of information transported in the
SOAP header, i.e. XML Encryption, XML Signature and security tokens,
which are being introduced in greater detail within this section. Tokens usually cover all the security related topics different from encryption or digital
signatures, such as SAML tokens for authentication or authorization. WS-
Security extending standards specify which tokens may be used, how they
are created, processed, transported and so on. WS-Security itself defines
username tokens for simple username/password authentication. Though not
very secure, it is a widespread authentication method and often sufficient,
depending on the application setting. Username tokens may come along with
plain text passwords or hashed passwords adding a timestamp and a nonce to
increase randomness and prevent guessing or replay attacks. Another option
would be to use shared secrets between service provider and invoking clients.
The only thing that needs to be taken care of is that both interacting parties
have sufficient information to repeat the hashing process and compare the
resulting password digest. Another form of security tokens are binary security tokens. Because they are binary, they need to be encoded, e.g. Base-64,
in order to be represented within XML documents. Up to now, WS-Security
specifies X.509 certificates, version 3, and Kerberos tickets. XML tokens are
XML formatted information, where each type defines its own root element,
such as SAML assertions, XrML licenses or XCBF (XML Common Biometric Format) [121] tokens. SAML and XrML will be discussed shortly.

To inform Web service consumers about the security requirements a Web
Service has, WS-Policy [16] provides a framework to express interaction rules
in a machine readable manner. WS-Policies may either be associated with

a set of Web services by letting a policy point to the Web services it applies to or referencing a set of required policies within a WSDL document. This differentiation is important, since Web service consumers can but not necessarily need to be Web services and thus may not be described by a WSDL file, though policies may also apply to clients. WS-Policy seems like the perfect endorsement to WSDL, which was always criticized of being too static and not offering functional or non-functional service documentation. WS-Policy may introduce more dynamics to Web service descriptions and could even be suitable for policy negotiation among a set of Web services and interacting clients, which was long discussed in context with agent technology, or a useful support when handling Web service evolution [138]. More particularly, WS-Policy defines a framework for specifying policy assertions (WS-PolicyAssertions) [28] and how these resources are bound to a specific resource, for example WSDL (WS-PolicyAttachment) [17]. WS-Policy is applicable to a wide number of different functional domains, such as reliable messaging or Quality of Service (QoS) characteristics. An assertion must specify whether it must, must not or may be applied. Additionally, requestors may be informed as soon an assertion is applied or ignored. An optional preference number indicates the priority of the policy assertion. Multiple assertions may be combined using the `<All/>`, `<ExactlyOne/>` or `<OneOrMore/>` element operators, where `<All/>` is the default value.

WS-SecurityPolicy [57] builds upon the set of WS-Policy recommendations. It embodies a policy description language specific to expressing security constraints of a Web service, such as required security tokens, encryption algorithms or signature processing. This is exceptionally hard when we remember the complexity involved standards have to offer. WS-SecurityPolicy currently supports username tokens, XrML licenses, SAML assertions, X.509 certificates and Kerberos version 5 Ticket Granting Ticket (TGT) and Service Ticket (ST). Token issuers contain the names of trusted issuers. Claims further describe the type of token in use. This affects username, X.509v3 and Kerberos token types. Username fields need not match exactly, prefix values and regular expressions are allowed as well. According to WS-Security, passwords may again be specified in plain text or as hashed digest value. X.509 extensions allow adding more descriptive data to the certificate. Integrity determines whether a SOAP message needs to be signed, which parts need to be signed and which algorithms to use for signing. Usage and preference attributes may be declared as discussed with WS-Policy. Allowed algorithm types are canonicalization, signature, transform and digest algorithms. Algorithms may further be specified by adding algorithm specific elements and attributes. Multiple algorithms of the same type introduce more flexibility and allow the other party to choose which one to use. Security tokens and claims as mentioned previously may also be used. Which parts actually need signing may be indicated by XPath 1.0 expressions or functions defined by the WS-PolicyAssertions specification. Confidentiality defines which parts of the message need to be encrypted. It supports exactly the same features as integrity does, except the claim concept and the only supported algorithm type is encryption. Visibility configures the SOAP message parts which must be visible to intermediary message recipients, such as routers or firewalls. Visibility refers to either plain text message portions or messages encrypted in a way that intermediaries can handle them, i.e. using keys any intermediary must know. Additional assertions include security header, which allows constraints to be put on the message security header and message age, which is similar to the WS-Security timestamp.

We saw that WS-Security is extremely powerful in terms of expressing security on message level. We also saw that such dimensions of complexity have their price. The vast number of configuration options Web service specifications have to offer may result in huge message sizes and thus serious performance bottlenecks. WS-SecureConversation [3], developed by IBM, Microsoft, RSA and VeriSign tries to solve for Web services on the messaging layer what SSL does on the transport layer. It establishes a mutually authenticated security context, which can be used to transport a series of subsequent messages. This significantly reduces the size of the message headers, and overcomes inefficiency that equal messages have to go through the same security checking process, which is computationally expensive. Like SSL, WS-SecureConversation uses public key cryptography to exchange a secret key, which is used for encrypting and decrypting SOAP messages. The so called security context token contains an identifier, a created and expiry timestamp and an optional list of secret keys. If no secret keys are specified, it is assumed they are already known between the communication partners. Security contexts may be established by letting a third party create the security context token or one of the communication partners is responsible for it or by negotiation. Token exchange is performed through protocols defined by WS-SecureConversation and WS-Trust.

Trust is a difficult issue, which raises many questions about whom to trust and under which circumstances. Each communicating party has to decide whether it can trust the credentials provided by the other party. Hence, WS-Trust [4] needs to formulate methods for exchanging trust related security tokens and establishing trust relationships as extensions to the underlying WS-Security layer. WS-Trust tokens are similar to SAML request/response tokens discussed later on. A Security Token Service (STS) serves as trust broker between communicating parties of probably different trust domains. A Web service notifies the invoking party of eventually present security policies. The client may either already possess them or contact the STS for retrieving the required token set. This model might become arbitrarily complex, since each STS could be provided as secured Web service with own policies as well.

WS-Privacy is a not yet published standard, which may be compared to the W3C Platform for Project Privacy Preferences (P3P) [171]. Designed for the Web application domain, P3P allowed a server to publish privacy preferences, which compared with the user preferences as specified in the Web browser options, notified the user of privacy conflicts.

Web services may often have different security requirements as described by WS-Policy. To establish a common security context among a set of interacting Web services, WS-Federation [102] aims at translating one security token into another via Secure Token Services (STS). If a client wishes to access a Web service, it requests a security token from its STS. With this token, the client may contact the service STS and obtain the tokens required for interacting with the service provider. The only precondition in this setting is that both STS entities have established a trust relationship between each other.

WS-Authorization [132] has not yet been published but is very likely to aim at tailoring the concepts introduced by SAML and XACML to Web service environments.

2.4.4 Related Specifications

The Security Assertion Markup Language (SAML) [114] is a standardized XML format for describing identities and assertions identities want to make in a manner portable across company boundaries and also support secure distributed transactions. Features include suitability for any underlying transport protocol, it is XML and thus any available XML processing tools also apply to SAML, it is a standard message exchange protocol and does not require a central certification authority, since security is expressed in form of assertions.

Although SAML predates Web services and related security standardization efforts, it exactly addresses the needs of Web services for portable identity, which led OASIS to integrate SAML 1.1 into the WS-Security layer. Web services aim at cross-domain usage and thus demand portable trust specified by WS-Security. The problem is that SOAP supports no means to communicate security properties to establish trust relationships. Collaborative commerce environments however, need all participants authenticated in a compatible way across the entire virtual system spanning multiple trusted domains. SAML provides a common identity infrastructure. It facilitates building cross-domain trust enabled contractual agreements, which provide single sign-on while allowing each participating entity using its own authentication system and thus retaining the loosely coupled nature of Web services to establish B2B integrations. In traditional information systems the user provides its credential, for instance as password, which is compared with the permissions stored in the company's information systems. In other words, each company hosts its own individual trust domain, which is too restrictive for interoperation among company boundaries. Furthermore, in complex organizations, a large number of users and applications must be managed, each probably owning multiple accounts. A solution to this problem is single sign-on, enabling a user once logged in, to having access to all services contained within a trusted domain via cross-domain contractual agreements.

SAML specifies to carry assertion information, including authentication and authorization attributes, within SOAP headers. Security expressed in the form of assertions about subjects, which are credentials used to initiate some action. These assertions as well as protocol request/response messages and protocol binding information are specified via XML schemas. Hence, interoperability is given at the specification level. An assertion contains information about the claims an issuer is stating, including validity conditions, authentication method and subject identifier. There are different types of assertions for various purposes: Attribute assertions certify that a particular subject has particular attributes/properties, authentication assertions certify that a particular subject was authenticated at a particular time and authorization assertions can be used to transport access control decisions and/or authorization polices. Authentication is performed by a trusted third party authentication authority, which evaluates the credentials provided by the requestor according to access control policies. It subsequently creates an authentication statement containing the authenticated subject, i.e. the requestor, the authentication method and the authentication timestamp. Authentication is given when the authentication authority has authorized the subject of an assertion at a particular time with validity for a certain period of time. In other words, the subject is who it claims to be and prove that credentials are legitimately in the possession of the subject. SAML supports password-based authentication as well as Kerberos tickets, secure remote passwords, hardware tokens, SSL client certificates, XML digital signatures and public key authentication following the X.509, PGP, SPKI or XKMS key model. Analogous, an authorization authority replies with an authorization statement containing assertions that enable a subject to access a resource for an assured amount of time given certain evidence. SAML architecture comprises a Policy Decision Point (PDP) making decisions about access control given a set of parameters. A PDP may have access to an external Policy Retrieval Point (PRP), which delivers the policies required for the decision making process. Each time a decision is being made the Policy Enforcement Point (PEP) is called, which in turn forwards the request to the appropriate PDP for decision making. The SAML standard has been implemented by OpenSAML efforts [90] and is also integrated in Liberty project, which is comparable to federated identity management as achieved by Microsoft Passport.

Known originally as XML Access Control Language (XACL) [81], the eXtensible Access Control Markup Language (XACML) [103] was approved as OASIS standard in 2003. XACML builds on SAML and is an extremely powerful language for describing general, fine-grained access control for all kinds of computing platforms, handling policy enforcement at any point in the system and allowing different access control enforcement mechanisms. On top of the XACML data model is a policy or a set of policies, which may contain policies or recursively other policy sets. Policies are combined via reusable rules, which evaluate to Boolean expressions indicating either permit or deny. Consequences of rules are called effects. Targets define sets of resources, subjects and actions, a rule applies to. Obligations are operations defined by policies, which must be performed in context with the enforcement of an authorization decision. Decision making in XACML is actually done via combining algorithms, which may be used by both policies and rules to build up increasingly complex policy hierarchies.

The XML Key Management Working Group [69] works on a general concept on key management. A first XML Key Management Specification (XKMS) [83] was published in 2001 and was applied for example by Verisign. In the mean time a new proposal XKMS2 was published in 2004. Currently, interoperability tests between prototype implementations are conducted. XKMS was defined on top of XML Encryption and XML Signature to provide a technology similar to Public Key Infrastructure (PKI) as trusted service for Web service consumers and providers. PKI handles distribution, certification and life cycle management of cryptographic keys. Since PKI has proven to be very expensive regarding maintenance in practice, XKMS relieves Web

service requestors and providers of having to build an own PKI. XKMS specifies an architecture for generation and registration of private and public key
pairs. This architecture addresses also the mediation between existing approaches, such as PGP, SPKI or X.509. XKMS is divided into the XML
Key Information Service Specification (X-KISS) and the XML Key Registration Service Specification (X-KRSS). X-KISS is responsible for locating
and validating key information, which is being used by communication endpoints without direct access to required security information. X-KRSS offers
functions for registering a public key at a trusted registration server, key
recovery, revocation and reissuing, which may be applied in cases common
X-KISS services are not sufficient. A typical application scenario might be a
Web service receiving a certificate it cannot process. In turn the Web service
would forward the certificate to an X-KISS Locate service and receive a key
value. Validation would be performed in connection with an underlying PKI
service. The server interface abstracts from existing PKI approaches and is
implemented by means of Web services.

The eXtensible Rights Markup Language (XrML) [184] is used specifically
to control access to digital content using arbitrarily complex access rights
definitions. The data model defines a `Principal` who is responsible for pre-
senting the required credentials, a `Right` object an abstract `Resource` to
which a principal may gain access and a `Condition`, specifying the terms,
conditions and obligations, which must be met before accessing a resource.
The root element of XrML is a license, which wraps principal, right, resource
and condition into a grant object.

2.5   Access Control Approaches

2.5.1   Traditional Access Control

Access control has a long history in information systems. Multi-user operating systems require access control to avoid conflicts between users. Database
servers or Web server software incorporate a more sophisticated control model,
since such servers are intended to distribute information to different users.
Access control is typically specified by assertions stating that a certain subject has a certain privilege to perform an action on a specified object. Subject
and object may also be groups of subjects or objects respectively. Access control ensures that access is granted only if the requesting subject is entitled
to perform the requested operation. Access control is a fuzzy term, which
is often used in the context of authorization, policy enforcement or decision-

making. Well, these concepts are very closely related to each other and are often used in an interchangeable way nevertheless there are subtle differences, which will be discussed within this section. Access control is about verifying access rights, whereas authorization is about granting access rights [159]. Controlling access is basically about protecting objects from unauthorized access of subjects. Subjects are very often processes acting on behalf of a user or independently to perform a certain action.

A common approach to oppose subjects and their access rights on objects is by an Access Control Matrix (ACM), where each subject is represented as row in a matrix, each object as a column and cells contain the allowed actions. Since many cells may be empty, an ACM is often not the right choice to store permissions efficiently. An Access Control List (ACL) stores only the permissions that may be actually granted as triple of subject, object and access right and omits empty entries, indicating that no such permissions exist. ACLs represent the ACM by column and are attached to objects. Capability lists are similar to ACLs but represent an ACM by row and are assigned to subjects. Since ACLs can grow very large themselves, an alternative approach is to use so-called protection domains. Protection domains hold a set of object and access rights pairs indicating which action may be carried out on an object in a certain domain. When a subject is authenticated it can be determined which protection domains it belongs to and which actions it is entitled to carry out.

This approach is strongly related to the concept of Role-Based Access Control (RBAC), which organizes individual users into groups to facilitate the administration of access control policies. Instead of browsing through the permissions for each user registered in an information system, it is only required to check, whether a user belongs to a role which is allowed to perform the desired action. A simple example is the case of a company's intranet, which each employee may access, but external users must be prevented from gaining access to the system. It is sufficient to test whether the current user is member of the employee role and grant access accordingly. In other words, roles determine the protection domains a subject may operate within. Hierarchical grouping of user roles can further improve efficiency, since permissions may be inherited starting from the up-most role down the tree hierarchy. The same applies to objects whenever they may be somehow interrelated with each other. Access control may be performed on many different levels in the information system architecture. Concrete implementation depends on the actual security needs. Strembeck and Neumann [154] present a general framework based on RBAC constraints to make authorization decisions using context information. An engineering process for context constraints is presented and describes design and implementation of an RBAC service enabling context constraints enforcement. Means for the definition and enforcement of fine-grained context-dependent access control policies are discussed as well.

Access control is often implemented based on an abstract architecture consisting of a Policy Repository, holding policy rules and policy related data, a Policy Decision Point (PDP), which represents a logical system which makes policy decisions, and a Policy Enforcement Point (PEP), which is responsible for enforcing policy decisions. Usually access control is performed as follows:

1. Client sends access request to PEP

2. PEP asks PDP to check access

3. PDP fetches policies from the Repository, makes and returns the policy decision

4. PEP enforces the decision and returns to the Client

Access control models provide an abstract framework for the definition of authorization policies. As mentioned previously, RBAC assigns permissions to roles and roles in turn are assigned to subjects. Roles represent responsibilities and user profiles in an organization. Roles abstract from individual users and facilitate the administration of authorization policies. At any time permissions can be assigned to or revoked from roles and roles can be assigned to or revoked from subjects. RBAC is the de facto standard for access control in software based systems and can emulate other access control models, such as DAC and MAC. Discretionary Access Control (DAC) allows the owner of an object to determine who may access the object and how it may be accessed. Permissions are assigned to individual subjects and/or to user groups at the object owner's discretion. Mandatory Access Control (MAC) aims to control information flow in a lattice of security classes to protect confidentiality or integrity of sensitive data. Such a lattice structure can be imagined as a directed graph allowing information flow from less secured protection domains to the most sensitive ones but not the other way round. Development of MAC based systems was mostly was driven by the military sector and is also applied in the SE Linux operating system.

Tolone et al. [164] review access control in collaborative systems. They distinguish a traditional matrix access control model, RBAC, Task-Based Access Control (TBAC), Team-Based Access Control (TMAC), spatial access control and context-aware access control. In a traditional access control model for each subject and object an individual assertion is required. To facilitate the representation a matrix is used to specify for each object which users may access it. In RBAC several users may be grouped and one assertion is sufficient to give a group of users certain rights. In a TBAC the context in which a user wants to access an object is recognized and he may only have access if the object is required for the current task. Thus the access rights are time dependent. The team-based approach seems especially interesting for virtual teams. In contrast to a role-based approach the grouping of users to a team is temporarily. Thus TMAC is also an approach to incorporate the context of the subjects. Tolone et al. describe further models that use contextual information to decide permissions.

2.5.2 XML Access Control

Different approaches address access control on the file-system level, XML however, represents an important opportunity to provide more fine-grained access control. Damiani et al. [54] propose an improved model for specifying the subject of an access control assertion. They claim that for E-Commerce and similar fields of application a file based specification as it is used in operating systems or current Web servers is insufficient. Hence, they use meta-information typically supplied by XML Schema [152] or Document Type Definitions (DTD) [173] to specify sub-trees of XML structures as subject of an access control assertion. Their model is applied for Web servers where typically read operations are performed and they give an algorithm how to restrict the view on semi-structured data when a user has not the right to read the complete structure. Further research is presented in [56]. Different degrees of sensitivity and needs for sharing portions of XML data demand models and mechanisms for the enforcement of access control policies on XML documents. [24] defines a formal model of access control policies and a mechanism for encrypting different portions of the same document according to different encryption keys, and selectively distributing these keys to the various users according to the access control policies.

The widespread use of XML demands the need for flexible access control models for XML documents to protect sensitive and valuable information from unauthorized access. [153] presents a novel declarative access control model and introduces the Xplorer engine for search-browse-navigate activities on XML repositories, which is also capable of auto-generating an access control enabled Web application according to access control rules. Access control on the basis of data location or value in an XML document is essential to restrict access to sensitive information. However, current approaches regarding

XML access control tend to work on individual documents and suffer from
the lack of scalability. [128] proposes the notion of Policy Matching Tree
(PMT), which performs accessibility checks and is shared by all documents
of the same document type. Goel et al. [74] use relationships between a set
of different XML documents. Access control rules are derived from schema-
level rules, document or database content using the XQuery language [25].
[36] focuses on usage control for XML documents and supports expressions
of access restrictions directly on XML elements and attributes as well as
data-types and reuse relationships between documents. Wang and Osborn
[174] analyze combinations of different access modes. The authors propose a
RBAC model for XML databases, which allows complex authorization models and propagation of permissions. The drawback of the approach presented
is that only roles and no single users are supported. This could be overcome
by defining roles on a much finer level of granularity by introducing role Ids.
For example members of role Employee could be assigned an employee Id,
such that a permission can be granted only to specific role members. [53] proposes access restrictions for XML documents that contains encrypted regions.

Fundulaki and Marx [72] give an extensive overview of most popular approaches using XPath for XML access control, such as XACL [81] or the
work done by Bertino and Ferrari [24]. The paper formalizes an specification
language for XML access control polices, including granularity, conflict reso-
lution and query evaluation. Murata et al. [117] introduce a static analysis
approach on node level and restrict the usage of XPath axes to forward navigation. Similarly Luo et al. [105] suggest primitive, pre-processing and post-
processing access control approaches with special focus on pre-processing
using Non-deterministic Finite Automata (NFA) to rewrite a query to elim-
inate any parts violating the access control rules. Static analysis using tree
automata is also subject to research efforts made by Yagi et al. [185]. Most
of these approaches only focus on authorizing read requests. Lim et al. [100]
also consider update operations in case users attempt to change the structure
of an existing XML document. The focus of [53] is how to control access to
XML documents, once they have been received. Access control policies for
restricting access to XML documents can be enforced by encrypting regions
of the document specified using XPath filters. Methods for minimizing the
number of keys distributed to users and comparisons to other access control
frameworks are subject to the paper as well.

So far, the role of XPath in authorizing queries on XML data has been
discussed. How this is actually enforced is subject to this section. XPath
is a convenient method for identifying portions of XML data. With access

control the role of XPath is twofold: for querying and protecting sensitive data. What is needed is an algorithm, which determines whether a permission that satisfies a query has been defined. Therefore two XPath fragments, the protected path and the query itself, have to be compared with respect to containment and equivalence. A series of research has been conducted in this field, such as a by Miklau and Suciu [112]. As already presented in previous sections, XPath incorporates many different language features, for example axis navigation, branching, i.e. filtering, or wildcard selections. Hammerschmidt et al. [84] deal with the intersection of XPath expressions and prove that the problem becomes even NP-complete if the negation operator is introduced as well. Schwentick [151] discusses different XPath features and provides complexity analysis of their combinations. Prior work has shown that containment and equivalence algorithms considering combinations of XPath features are coNP-complete. The authors therefore present a sound algorithm, which runs in EXPTIME and show that special cases are can be decided in PTIME. Although in some other cases the algorithm is again coNP-complete, a modification of the algorithm can efficiently handle those cases but sometimes may occur to return false negatives.

2.5.3 Semantic Access Control

The foundations of Semantic Access Control (SAC) are based on the Semantic Web architecture as illustrated in Figure 2.3 [21]: The SAC model is based on the semantic properties of the resources to be controlled, properties of the clients that request access to them and semantics about the context and the attribute certificates trusted by the access control system [186]. Access control should be location transparent indicating authorization can be performed independent of resource location. To positively identify a user or client, the credentials provided must be digitally signed as shown in Figure 2.3. The advantage of SAC is characterized by its flexibility and support for interoperability of authorization mechanisms in distributed and dynamic systems with heterogeneous security requirements. It enables the semantic validation of the access control criteria and promises simplicity, correction and safety of the system. [187] discusses a validation algorithm for detecting semantically incomplete or incorrect access control policies. Furthermore, a formal SAC model is presented along with some proofs of the model, which forms the foundation of the semantic validation algorithm. [189] introduces semantic layers to authorization components. The paper describes the development of an application framework, which incorporates a Semantic Policy Language (SPL) for the description of access criteria. The same approach is applied to Web service environments as illustrated in [188]. [190] presents an

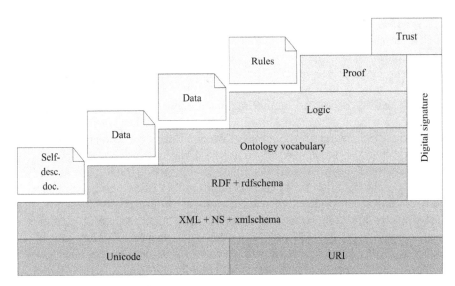

Figure 2.3: W3C Semantic Web architecture

access control system for Web services. SPL is chosen for the description of access control criteria based on the use of attribute certificates. SPL has been designed to take advantage of semantic information about resources and the context to achieve policy validation and facilitate security management. SPL applies traditional concepts of modularity, parameterization and abstraction in order to provide simplicity and flexibility to the difficult task of specifying access control criteria. The modular definition of SPL policies implies the separation of specification in three parts: access control criteria, allocation of policies to resources and semantic information, i.e. properties about resources and context. Additionally, SPL makes possible the abstraction of access control components and, as a consequence, the ability to reuse these access control components. All the previous properties help the reduction of the complexity of management. Moreover, the use of semantic information about the context allows the security administrator to include relevant contextual considerations in a transparent manner, also helping the semantic validation task.

In [129], Qin and Atluri develop a concept-level access control model for the Semantic Web by specifying access control based on ontologies using OWL. They identify and categorize domain-independent relationships among concepts and propose propagation policies. How user requests are handled is discussed as well. This approach describes an access control model that can be used for specifying authorization over ontology concepts and enforcing them over concrete instances. Demonstration of how concept-level security polices can be represented in an OWL based access control language is presented as well. The concept-level model is considered especially suitable for the specification and administration of access control over semantically related document data even if they conform to different DTDs or use different tag names. Providing inter-organizational data access to users across company boundaries is traditionally solved by sharing metadata using federated and mediated databases, which may not be acceptable for certain organizations due to privacy concerns. [115] introduces a Toolkit for Privacy-preserving Access Control (PACT) without having to share sensitive metadata. Ontologies, ontology-mapping tables, role hierarchies and queries are stored encrypted to maximize privacy and confidentiality of semantic data, while enabling interoperability among heterogeneous databases. Semantic access control is incorporated using ontologies and semantically enhanced authorization tables. Development of effective mechanisms for manipulating access and version control has become a major research area. Updating a document may require modifications in associated access control policies and vice versa. [41] deals with the integration of these issues into a single framework, such that different document versions can be assigned different access authorizations, and uses RDF as unified representation of access control policies. Based on RDF ontologies, [5] presents a flexible and interchangeable XML access control model using view-based, declarative semantics.

2.6 Project SemCrypt

The work described was performed in the context of a research project funded by the Austrian FIT-IT and carried out at the E-Commerce Competence Center (EC3) in collaboration with the Data and Knowledge Engineering Group at the Johannes Kepler University of Linz, simply referred to as DKE Linz, and the EC3Networks GmbH as industrial partner. The project Sem-Crypt [177] explores techniques for processing queries and updates over encrypted XML documents stored at untrustworthy storage providers [134]. By performing encryption and decryption only on the client and not on the server, SemCrypt guarantees that neither the document structure nor the document contents are disclosed on the server. The chosen approach exploits the structural semantics of XML documents and uses standard, well-proven encryption techniques. SemCrypt has been launched to investigate different approaches of how XML data may be encrypted on element level while still providing the possibility to modify the elements without revealing the internal storage structures. Supporting software was designed to rely on open standards in order to provide an extensible and flexible framework for building trusted application environments that may also be easily integrated in existing electronic workflow structures.

Inspiration for the SemCrypt project is funded in related research work. The idea of service oriented databases and related security concerns are presented in [79]. Outsourced databases need at first be protected against theft and secondly protected from service providers, since they cannot be trusted as well. The approach taken focuses on relational databases and thus on execution of SQL statements over encrypted data and letting the client take care of decryption. [80] follows a similar approach of database outsourcing. Administration is done by service providers. Data management is provided as service guaranteeing privacy. [55] deals with querying encrypted databases that are hosted by insecure remote servers and selecting data without needing to disclose database content. An indexing mechanism is used to balance the trade-off between efficiency and privacy requirements using B+ data structures. An evaluation of vulnerability is also presented. The architecture introduced in [39] uses a Trusted Privacy Manager (TPM) entity, which takes care of key generation and management as well as data encryption to enforce user privacy requirements on outsourced data. The system proposed provides scalability, standards compliance and the enforcement of privacy requirements of both data owners and consumers. It makes use of non-standard encryption strategies and relies on a TPM in charge of data encryption and key delivering. Furthermore, strategies for key generation and management are presented. As illustrated in Figure 2.4, query and update processing are shared between client and server, where as much as possible is done at the server and encryption/decryption being performed only at the client. The semantic-based solution is orthogonal to encryption techniques employed and, thus, widely applicable and independent of general technological advances in encryption. Servers provide special storage and access structures for storing encrypted fragments of XML documents. Clients exploit these special storage and access structures according to the given document's structural semantics, which is known to them, but not to the server. With neither the document structure nor the document content being disclosed at the server, the server need not be trusted with respect to maintaining data privacy. Query and update statements, written as if against a plain XML document, are mapped by the client to corresponding access primitives against the encrypted XML fragments held at the server. The techniques are explained in more detail in [134]. Part of the SemCrypt

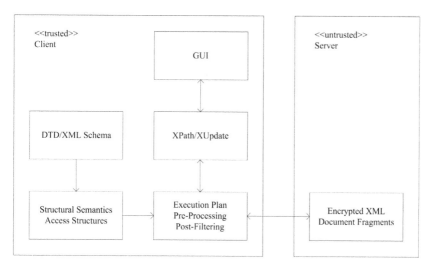

Figure 2.4: Initial SemCrypt approach

project is the development of an environment where the authorization of users and the encryption/decryption is takes place. This is especially required if the SemCrypt technology is to be used in applications with users having diverse privileges. The SemCrypt approach is evaluated in different applications. One of these applications is settled in the Human Resource Management (HRM) area discussed in Chapter 3.

Comparable database systems (DBS) transfer the complete data content to the client, decrypt the data and execute the query locally on the client machine. This leads to tremendous performance drawbacks due to huge data-set transfers. SemCrypt uses index structures, which are built upon available XML Schema information. These indices are used to efficiently query in encrypted data-sets [78]. An XML Schema is the starting point for each SemCrypt application, since all documents stored in the database must adhere to a single schema file. Chapter 7 provides a step-by-step documentation of how to develop and deploy a SemCrypt compliant application. The SemCrypt DBS consists of a database management system (DBMS) and a database server. The DBMS is moved to the trusted client domain and manages the data-sets, which is made available through the database server. The actual physical storage is done by a conventional relational DBMS (RDBMS).

The SemCrypt DBS is built of multiple layers providing different views on the storage data (2.4). The lowest layer represents the stored documents as key/-

Layer	View
External	XML
Logical	Tree-based
Internal	Path-based
Physical	Id/Value pairs

Table 2.4: Representation layers in the SemCrypt database system

value pairs, completely destroying the semantic structures of the XML data, to additionally leverage security measurements by making statistical analysis of the data storage almost impossible. For improved internal data processing, the internal and logical layers provide appropriate data structures. XML has distinguished as emerging standard for semi-structured data. Therefore, XML is used for representing the database content to the end-user, which may then be further processed using XSL and related technologies.

Part II

Concepts

Chapter 3

Use Case Scenarios

The following chapter concentrates on the identification and specification of possible future application settings applying the SemCrypt framework technology, aiming at supporting both single-user and multi-user scenarios. Beginning with a few considerations about single-user deployments, the main focus of this chapter lies on the in-depth analysis of selected large real-life application areas. Thereby the applicability of existing XML technologies is crucial. Criteria for use case selection are ostensibly high data sensitivity, collaborative query and update operations, applicability of XML for interoperability, existing, adoption of XML as communication protocol and storage format, XML initiatives, acceptance and demand, user role models, importance of archiving and securing large data sets, legal aspects with respect to privacy and data protection, relevance for ICT and market potential of security applications.

Very often application settings are something in between multi- and single-user. Single user as well as multi-user environments may both occur in centralized and distributed applications. The following table opposes application examples classified by computing organization and user interaction. Since we can assume that private document data stored locally on a user's private machine, which is not even connected to the Internet, is safe as long as it is not accessed by someone who is not entitled to. Additionally, file encryption tools provide an additional notion of security. SemCrypt however aims at

	Single-User	Multi-User (Collaborative)
Distributed	Web Hosting	Human Resource Management
Centralized	XML Spy	Multiplayer Games (Offline)

Table 3.1: Classification of application scenarios

far more challenging security management in distributed systems and thus concentrates on the left column of the table 3.1 presented above. Widespread applications, such as file or Web hosting services can be seen as distinct single-user applications, since they do not imply any notion of collaboration among the users accessing the service. Files are stored at a remote server and belong to exactly one user who has full access to all private document data. The security model here is quite simple: access control is based on the outcome of the user authentication. The challenge in this case is to develop encryption strategies that protect remote data from unauthorized access. Based on the fundamental architecture design presented at the end of Chapter 2, it is up to the application designer to choose whether to provide a trusted domain or not. Subject to protection is the cryptographic key, which both must be applied on the sensitive data in order to get useful information out of them. Both may be stored directly or detached at the user side or at the trusted domain as long protection from unauthorized access can be guaranteed. Apart from key management, further design decisions comprise whether the user or the trusted application provider is responsible for key generation and how to accomplish dynamic key change during application runtime.

Whether data should be encrypted or not also is dependent on the design decisions of the application developers. Sometimes, it might even be undesirable to store data encrypted in case the data provider wants to analyze the database. This is happening with Google services, such as the Mailing or Calendar services, for instance, which index client data and extract information out of it for scientific reasons. Access control, however, is indispensable in SemCrypt environments. Chapter 5 gives an in-depth analysis of authorization processing in SemCrypt applications and provides a specification of the access control definition language in use. In each application setting, at least one access control policy must be used, which is valid for all documents and users registered in the system. In some applications, users may wish to introduce own access control rules for their private documents. Hence, permission and denial inheritance and overriding is a fundamental feature being supported by the SemCrypt environment.

The following sections contain an analysis of different application domains, in which the application of SemCrypt technology is considered reasonable. Each section provides a general introduction of the topic and demonstrates its usability in the context of single- and multi-user application settings. These scenarios have been identified in the course of the SemCrypt project for demonstration purposes of the technologies developed and have been made publicly available for download on the project Web site [177].

3.1   Human Resource Management

Human resources (HR) are in today's organizations the most important resource for running the business and for continuous innovation of business
models, processes and systems. Typical Human Resources Management
(HRM) tasks are the development of an organization's workforce and the
staffing of an organization [180]. Another important task in HRM is the
assignment of work force to given tasks. If an inter-organizational project
is started, different human resources are assigned to the project. The assignment has to consider the skills of the people, the interests of individual
partners as well as those of the employees of the partners, the availability of
the personnel as well as costs attributed to such resources. Hence, to assign
human resources to a project, different attributes of employees of different
business partners must be accessible. However, such data about personal
attributes is highly sensitive and must only be accessed in an authorized
manner. In some situations access over the boundaries of a single company
may improve the flexibility of a virtual organization, will speed up business
processes and improve the quality of decisions. Thus, a sophisticated access
control management is required in virtual enterprises. We may restrict the
access to certain roles and assign people to such a role in order to give only
few people the possibility to access certain artifacts. For instance, in each
participating organization we may identify a person who is responsible for
the cooperation with other partners. One problem that occurs is how to
specify the artifacts that a certain person, role or process may access. There
may be information that is made accessible to any public audience such as
marketing information, there is information accessible to business partners,
but employees may access different information and finally there may be information accessible only by certain members of the staff.

The field of HRM offers a wide range of possible application scenarios. In particular HR Outsourcing (HRO), as it is already practiced in the United States
and is very likely to being established in European countries as well, lends
itself as prototype environment for testing and demonstrating the achievements made in the course of the thesis for the following reasons:

- Extensive query and update requirements for HR data

- High sensitivity of covered information due to legal regulations, competitive advantage and privacy aspects

- Adequate amount of sensitive and non-sensitive data

- Availability of standardized vocabulary for data exchange (HR-XML)

- Multi-user scenarios: different access rights for different user roles

Furthermore, HRM as horizontal application branch is involved in many areas including E-Government, E-Health, E-Tourism, insurance, finance and production. On first sight, HR settings can be subdivided into inter-organizational scenarios covering information flows between companies and intra-organizational applications deployed within one single company. The ideal scenario would contain features of both aspects, such as external recruiting services and internal staffing requirements. An example for a hybrid form of both approaches would be a project, which involves multiple companies and requires external as well as internal manpower for successful completion. External acquisition is performed via an external recruiting partner. For purposes of flexibility demonstration, alternatives to the primary search result regarding qualifications or salary could be displayed as well. Thereby it would be possible to query for qualitative differences of the employees and support employers in the decision process whether to employ additional specialists or provide additional training courses. Drop outs during project runtime could be compensated and changes in project duration or project aims could be bypassed. Background checks including verification of the provided qualification information or relationships to other companies could be performed automatically. After project termination, update and archiving of a possibly large amount of XML documents is necessary, involving staff qualifications, publications, project report or financial services.

In Chapter 7, a simple HRM application is documented, which allows students to upload their private resumes and competencies. Each student is allowed to edit and query his own documents, which demonstrates the usage of HR material in multi-user environments. Multiple roles, such as HR managers and teaching staff, have been introduced, with each role having different access rights on the students' documents. A student, of course has full access to his own documents, other parties have limited read access. Furthermore, a student may consider the access permissions of the application environment as not suitable for his private documents and thus upload a user-defined access control policy to override the default permissions and/or denials provided by the application server.

The HR-XML Consortium [86] is a non-profit group developing standard vocabularies and XML schemes for the HR domain. Members of the consortium are companies offering HRM software, recruiter and personnel consultants. HRM encompasses a diverse range of business processes such as advertising open positions, enrolling employees and their dependents within benefit plans, and ensuring that changes in employee status are recorded appropriately in internal information systems as well as the systems of external partners and service providers [58]. There are XML schemes for different HR processes as well as core schemes used to represent different aspects of HR, such as contact information, payment and skills or competencies of these resources. A sample schema incorporating parts of the HR-XML standard can be found in Appendix A and is used as reference scenario within the SemCrypt project introduced in Chapter 2 as well as for testing purpose of the approaches developed in the course of the thesis. This schema is as well used by the previously mentioned student scenario, which is further discussed in Chapter 7. A very simple but valid sample document according to the schema file may look as follows:

Listing 3.1: A sample SemCrypt employee description

```xml
<?xml version="1.0" encoding="UTF-8"?>
<SemCryptEmployeeInfo xmlns="http://ns.hr-xml.org/2004-08-02"
    xmlns:xsi="http://www.w3.org/2001/XMLSchema-instance"
    xsi:schemaLocation="http://ns.hr-xml.org/2004-08-02 HRMSimple.xsd ">
    <PersonInfo>
        <PersonId>
            <IdValue name="studentId">9902261</IdValue>
        </PersonId>
        <PersonName>
            <FormattedName>Wolfgang Schreiner</FormattedName>
        </PersonName>
        <ContactMethod>
            <InternetEmailAddress>
                e9902261@student.tuwien.ac.at
            </InternetEmailAddress>
        </ContactMethod>
        <PersonDescriptors>
            <LegalIdentifiers>
                <Citizenship>Austria</Citizenship>
            </LegalIdentifiers>
            <DemographicDescriptors />
            <BiologicalDescriptors />
            <SupportingMaterials />
        </PersonDescriptors>
    </PersonInfo>
    <Competencies>
        <Competency name="Programming Language" required="true">
            <CompetencyId id="Java" idOwner="9902261" />
        </Competency>
    </Competencies>
    <JobPositionHistory>
        <JobHeader validFrom="2005-04-18">
            <JobId>
                <Id>EC3</Id>
            </JobId>
            <JobTitle>Researcher</JobTitle>
        </JobHeader>
    </JobPositionHistory>
    <!--
        <Salary />
        <AssessmentResult />
```

```
   -->
</SemCryptEmployeeInfo>
```

Root of the document is always the `SemCryptEmployeeInfo` element, which contains `PersonInfo`, `Competencies`, a `JobPositionHistory`, current `Salary` and an `AssessmentResult`. Latter elements are optional and are omitted due to their complexity and space limitations. HR-XML defines these types in separate schema files. Apart from name and address, personal data comprises contact methods, legal identifiers and biological and demographic descriptors. Competencies may be arbitrarily nested to define skills on increasingly fine-granular levels of specification. One major research interest in HRM and related field is knowledge and competency management, i.e. how to describe and evaluate personal skills. Job history, salary information and assessment result elements should speak for themselves and contain especially sensitive information, which have been omitted in this short example.

Typical queries and updates cover statements, such as:

- Project manager selects employee with certain competencies and salary expectations and who is available in the next few months

- Employer inserts new employee with personal data, qualification and availability

- Project manager books employees for a certain period

- Update salary, qualifications, experience, assessments, availability

The HR scenario demonstrates the tremendous significance for permissions for accessing sensitive data of varying importance for a number of different user roles. Applications allowing multiple users operating on the same data sets additionally require a versioning mechanism for XML document data. Document versioning might be also desirable for history recording. As an example, from company internal permissions, we could define a set of user roles (CEO, HR manager, project manager, employees), a set of paths uniquely identifying portions of XML data (person description, contact method, competencies) and access rights. Inter-organizational data exchange would be much less questionable when using standard security techniques. The situation is slightly different for a job applicant. A classic resume is usually less security critical, but also becomes subject to encryption as soon as personal information is provided extensively, such as biometric information.

3.2   Project Management

The following show case shall briefly discuss features and requirements of the deployment of SemCrypt technology in project management processes. In inter-organizational projects, documents are often exchanged between company departments. In order to facilitate system integration, documents are often described using semantic Web technologies, such as RDF [172] and Microsoft (MS) Excel is especially common in small and medium enterprises (SMEs) for project management tasks. Fine granular access control on XML formatted MS Excel documents is not a major objective in these environments. Unauthorized access is prohibited by locking documents and operating system folders per se. The big drawback of this approach is that access control is performed location dependent rather than on the sensitive content itself. However, to extract data structures from MS Excel tables is a necessity and daily practice in many SMEs. Since MS Excel tables may be compared with relational database tables, mapping XML document data to MS Excel table is a challenging task, but offers additional possibilities for the deployment of the MS Excel based project management approach.

The project management domain is in some way closely related to the human resource sector discussed in the previous section. It comparably involves HRM or controlling issues and many more and likewise has many different types of users interacting with each other, including HR managers, project leaders and the like. Possible tasks in a project management environment comprise:

- Collaborative management of project deliverables

- Management of confidential financial data or research results

- Document management, including resumes and project schedules

The show case presented in context with the SemCrypt technology evaluation relies on the standardized XML project management format PMXML [52]. Documents following the PMXML schema are imported into MS Excel creating resource, scheduling and controlling spreadsheets. Documents that are created during project lifecycle often contain sensitive information and underlie the Data Protection Act. Project members occupy different roles of responsibility in a project and have very different access rights regarding the project documentation. Since MS Office products are widely applied in many companies worldwide, MS Excel spreadsheet documents shall serve as starting point of the analysis. By applying SemCrypt technology and adding meta-data over actual XML formatted information, a general improvement of information management can be achieved. A demonstration of the SemCrypt framework in combination with widely applied and accepted technologies, as MS Office applications are, would on the other be a promising argument for the overall acceptance of the SemCrypt project. Further details can be found in Chapter 7 dealing with provisioning of the SemCrypt security services on the client side.

3.3 Document Archives

Another interesting setting of applications is located in the area of sensitive document archives and mobile devices. The challenge is to identify relevant meta-data, which may be for formulated using XML Schema, and classify their security requirements regarding access control and cryptography. In any use case typical query and update requests for each user or user-role on the encrypted storage provider must be identified as well to evaluate opportunities and limitations of the approach followed. Especially for mobile devices required resources of the security modules is a critical issue. In case of client-side encryption, hardware must be capable of performing the necessary steps. This analysis does not only apply to mobile phones. Resource availability issues also concern Palm, Blackberry and Pocket PC devices.

One specific application is that of encrypted E-Mail storages. E-Mail applications are typically single-user, since each user normally only has access to his private mails. Deployment of SemCrypt technologies on the one-hand ensure that data is encrypted and thus secured from unauthorized parties and on the other allows E-Mail indexing, enabling comprehensive search functionalities without the need to decrypt the whole E-Mail storage. This scenario has been developed as show case for single-user document management and is also briefly mentioned in Chapter 7. A concrete need for secure E-Mail services could be identified for private home users, who wish to backup their data on some remote storage provider, or companies, who do want to be sure that no unauthorized party gains access to their IMAP server. Although initially intended as single-user scenario, it could be conceivable to make a mail server open to multi-user access. Hence, users could define policies and allow other users (restricted) access to private E-Mail documents, which especially makes sense if an E-Mail has been accidentally sent to only one person while it was intended for the whole company department. The original sender would then not need to resend the message or need to know the addresses of all the recipients, but the mail could be made publicly available on the company's mail server.

For the E-Mail application, query and update statements could be:

- Select all E-Mails received last week with subject S

- Select all urgent E-Mails from sender S

- Select all E-Mails, which contain certain keywords in the E-Mail body

- Insert a new E-Mail

One multi-user application example is settled in the field of transport logistics, which is also discussed in Chapter 7 later on. A transport company may inform a lorry driver about new orders via SMS. An order may contain information about the type of packages, time and location. Afterwards the driver should be able to decide whether the delivery shall be carried out immediately or be stored on an external storage server. After completing a delivery, the driver removes the order from the server. Storage data could be used as record history in order to keep track of incoming and finished orders, to optimize a company's logistics department, for quality management and similar problems. Many similar applications could be found in different areas. In case of the logistics application, encryption may be performed on the client side, i.e. the driver's mobile device, or on the security server before writing the content of the order message to persistent storage.

Possible query and update statements involve:

- Driver selects destinations with distance D to the driver location

- Driver selects orders with a maximum of P packages

- Company updates order with new items or delivery address

- Driver removes finished delivery from data storage

A typical scenario would start off with an order to deliver packets from one location to the other. This order is then transmitted to a server and stored in the database backend. To notify the lorry driver, an SMS message is sent to his mobile phone. If the driver is not immediately capable of fulfilling the order, he may wish to store the order message on some external server, which may not be secured from unauthorized accesses. SemCrypt technology could be applied to ensure that no-one except the driver can access the order message. Fine-granular access control on SMS level may not be required in this specific setting, since an SMS can be considered as single document introducing some single-user and non-collaborational context. In this context, different configuration settings may be considered as well. SMS transfer may happen unencrypted or encrypted, while the order message must be encrypted before actually stored, whether on the client's mobile phone or on a remote security server. Since SMS messages have a limit of 160 characters in size, an optimization of these procedures could be to only send a reference number of the order to the driver's mobile phone, while storing detailed order information externally, the driver may access via the reference number. We could even simply publish new orders on the remote server and let the drivers decide, which order suits them most, depending on their current location and availability.

3.4 Tourism

Internet technologies have radically changed the tourism market in the past years and lead to an integration of available information and offered services. Online search engines query tourism and destination Web sites to assist vacationers in increasingly complex holiday planning, such as viewing available hotels and booking flight tickets or rental cars. As complexity grows, also IT support becomes sparse.

Tourism has always been a main area of interest for the MOVE research group [61]. One reference scenario was the packaging of different products to holiday packages. Due to its complexity many technical and organizational issues have to be taken into account. Holiday packages refer to the combination of a set of holiday services that together make up a complete holiday trip. There are four levels ranging from pure tourism information, simple products or from supplier- to consumer-defined holiday packages. Consumer-defined holiday packages build on the foundation of the concept of supplier-defined holiday packages. A system at this level enables the customer to pick a supplier-defined holiday package and adapt it even further to his personal needs thus constituting the first step towards consumer-defined holiday packages. Customization possibilities are price, departure time and length of stay. This conforms with a general trend from a supplier market to a consumer market where the demand side specifies which products are delivered. Besides searching and packaging, the management of customer data and the generation of customer profiles is an important task. By collecting data of customers and their classification, service providers may organize special services that address the actual requirements of customers.

A tourist should be given the possibility to define a personal profile containing information about contact, such as E-Mail, phone number or home address, authentication certificates and personal preferences regarding accommodation and recreational activities. Such a profile facilitates the creation of custom-tailored services. Such a scenario is highly dynamic, since it has to adapt to very special factors during runtime. Main requirement derived from this scenario comprise among others security, authentication and privacy the respect to protection of consumer data. If we consider services such as transporting luggage of a customer as well payment with credit card and much more, security is an important issue. Services defined in the virtual enterprise must be clearly distinguished whether high security is required or not. A related issue is authentication. If a customer asks for services such as booking a room an authentication is necessary. Privacy on the other hand deals with availability of consumer data to other entities involved in such a setting. For many tourist agencies it may be interesting do perform statistically evaluate data of their guests, which requires the affirmation of the data owner. Usually, a customer would not like to define a very fine-grained level of access control to their data. Therefore certain common permission definitions may be useful.

3.5   Biotechnologies and Human Medicine

XML has become the foundation of several markup languages for storing biological data and is particularly important for biological research and the biotech industry [178]. These markup languages support genetic and biological data collection, management, retrieval, analysis, exchange, and publication. The growth in data produced for research in the field of life sciences is extraordinary as well as the need for standardized exchange formats for distributed processing of data or research in general, which is commonly organized in networks. For example, data is accumulated in the area of the Human Genome Project [87], for various genetic and proteomic databases or specific applications in life sciences. Most of the data involved is linked to human beings, which leads to an aggregation of sensible information of individuals and therefore poses a privacy and security risk. At the current (research) stage, most of the data aggregated is eventually dedicated to publication with private information of probands detached from the data sets. However, the existing ways of dealing with data exchange and privacy provision are discontenting, circumstantial, and aggravate the inclusion of network partners in research projects. XML is broadly accepted as a promising means for tackling the before-mentioned problems in bioinformatics, but does not solve the privacy and security concerns. In recent years, the pharmaceutical industry has gradually altered its research policy. Instead of conducting research on their own, pharmaceutical groups are progressively trying to pass on the risk involved in research to biotech SME, which are on the forefront of research but notoriously lack of capital. The pharmaceutical groups support or buy the biotech SME only if the drug is successful in initial testing phases and rather concentrate on sales and distribution than on research itself. As a consequence, research in the field of life science is frequently conducted by biotech SME that face SME-typical restraints in capacity, in particular financial constraints that make it even more difficult to manage and store large data sets appropriately. Therefore, cost-efficient outsourcing of databases could be an interesting option, but poses a challenge as privacy and security issues need to be addressed as well as the access of partners in the research network to the data stored at a service provider. Even though data will finally be accessible to the general public, the pre-publication or pre-patent phase as well as the linkage of data sets with sensitive information exemplifies the need of a method that enables a secure and cost-efficient way of storing data, as conceived by the approaches of this thesis. The archiving of accomplished experiments seems to be a particularly rewarding field of application in this context.

3.6 E-Government

E-Government offers a wide range of services dealing with sensitive data that is usually stored centralized and archived for longer periods [179]. Therefore, issues of security and trust become prevalent, particularly when personal information can be gathered from the data sets, promoting the frequently mentioned "transparency of the individual" that is favored by information and communication technologies in general. Consequently, E-Government seems to be an interesting field of application, as long as XML-based data formats are involved in the E-Government services. These services can roughly be divided into categories Government-to-Government (G2G), Government-to-Business (G2B), and Government-to-Citizen (G2C).

Chapter 4

Architectural Requirements

4.1 Basic Architecture

SemCrypt applications may be basically distributed on three different domains, which are classified by the security level they require. A user in a SemCrypt application may store, remove, update or select data formatted as XML and held by an untrustworthy storage provider. Thereby users may operate on complete XML documents or only document fragments using the XPath and XUpdate languages, for instance. Since data is stored encrypted, a client must first contact the trusted security server and ask for authorization. If the requested operation is permitted and the user wants to perform an update operation, the key management, which is typically also located at the security server, is asked for data encryption before the data can be saved in the data store. In case of a query request, the data is fetched from the untrustworthy database and sent to the key management afterwards to perform data decryption. Essential domain interaction and entity distribution is depicted in illustration 4.1.

The elementary SemCrypt architecture only prescribes the division into three security domains. How an application is deployed in concrete depends on the concrete application requirements. In some cases, for instance, clients may not even trust the trusted domain and take care of key management themselves. In such scenarios, it is possible for the clients just to ask for authorization and perform the rest of the interaction process locally. For example PIN codes, which are required for the ATM, are directly stored on a small amount of memory on credit cards. If the card gets lost, the owner should immediately report this to the bank and lock the account. That is the reason why the client domain is rated as semi-trusted. An attacker may un-

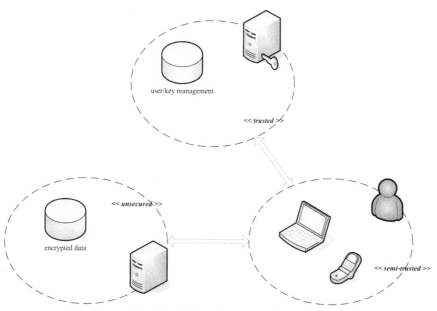

Figure 4.1: Basic SemCrypt application domains

der certain circumstances gain unauthorized access to the system, to which probability extent ever. Hence, if we manage to guarantee absolute data confidentiality, there is still a potential risk remaining at the client side. If we stay with the credit card scenario, the security server is physically moving towards the client domain. However, the authorization engine needs to be absolutely protected from any unauthorized modifications otherwise each client may change its access permissions. Also the data store must remain untrustworthy, since this is the actual challenge that needs to be solved in SemCrypt applications.

Figure 4.1 illustrates the simplest deployment setting for a SemCrypt application. Unfortunately, real world matters are often far more complex. SemCrypt basically aims at providing a database management system for handling outsourced sensitive data on stored at untrustworthy databases. A DBMS is primarily concerned with providing data persistence management in distributed applications, which makes the approach widely applicable in applications, where large amounts of sensitive data are involved and for which data outsourcing is more than just an option. Therefore it may likely be the case that data needs to be scattered on multiple data stores, which must to be queried, and several hundred client requests need to be handled at once, which forces the usage of multiple security services to meet the high performance requirements. Load sharing plays a very important role in these considerations and involves deployment of multiple trusted and untrustworthy domains, which need to interact with each other, also with respect to non-functional quality of service application requirements, such as service availability and reliability. Single clients do not require fine-granular access control to single XML nodes, since each client has full access to its private data. An example for single client settings is a Web hosting provider or a private document store, more generally spoken. Authorization in multiple client scenarios is dependent on the type of application. In collaborative environments it may be a requirement that all clients have full access to the resources, while in high security applications very fine-granular access control is demanded. Concurrent data modifications are not part of SemCrypt. For locking resources standard techniques are applied.

The component assembly and interaction document provided by the DKE Linz [93] provides an in depth analysis of the basic requirements of the security server interface and discussed different deployment scenarios. However, goal of this document is to give an overview of the components involved in SemCrypt domain interactions and introduces an application framework, which was designed to facilitate initial application development and deployment.

4.2   Responsibilities of Domain Entities

The storage provider actually contains encrypted data, only authorized entities can access in a meaningful way. Components required for decrypting database content are separated from the storage provider, which makes it cumbersome for potential storage hijackers to read the database content, since they do not have access to the required security data. Security of the data stored actually depends on the security level of the encryption mechanism applied. The stronger the encryption algorithm used, the harder attackers may gain access to the data content. The storage provider itself represents a physical entity not providing any application functionality and needs thus not be further investigated by SemCrypt technology. The challenging tasks are responsibilities of the security server introduced below.

Information contained by the storage provider is actually useless. The security server contains meta-data, which is required to decrypt encrypted sensitive information and return it to the requesting client party. Responsibilities of the security server are manifold and at the same flexible with respect to the management of the security data. We briefly saw that the security server is not an isolated server component by itself, but may rather be deployed with many different facets, sharing security functionalities with the client application, and sometimes be even physically moved to the client device. Functionalities comprise key management to store cryptographic keys for encrypting and decrypting sensitive data stored at the storage provider, request authorization, tightly coupled with authentication management of the subjects involved in an application setting. Authorization and access control needs to be protected by all means and must never allow any entity directly operating on access control policies. Key management is a little more flexible and may be shared with the client applications. Basically, the security server needs to be trusted. If trust in key management cannot be granted any application deployment becomes obsolete, since then it can also not be granted for access control, which is crucially important. However, in some applications, a client may need direct access to its keys without having to ask an external entity. In this case the security server only needs to perform authentication and authorization and return encrypted data back to the calling client, which performs data decryption on its own. Both access control and key management are subject to later chapters, which discuss the architecture and components involved in greater detail. The security server was designed to be applicable in any application without deployment limitation as regards the configuration options of the application components. Furthermore, every component provides a well-defined Web service interface incorporating service orientation while providing security with respect to identity management, access control and data protection.

Application integration should actually happen on the client side. The term client not only refers to a nice-looking, clickable graphical user interface, but refers to any application component that may wish to take advantage of SemCrypt technology. This engages a wide range of front-ends reaching from simple command-line tools for operating directly on the database to complex workflow management systems, using the security server as underlying persistency layer. Application purposes are far reaching demanding high configurability of the security technologies used in order to support the many needs that may be requested from such a security system. This very strongly argues for the conceptualization of the security server as Web service oriented architecture. It must be easily integrate-able into existing systems and stay independent from client platform and hardware. In the previous chapter we saw applications settings, which very well suit the functionality developed within this thesis and which already serve as prototype environments for applications that have been developed on the foundation of the technologies presented herein. These applications will be discussed later on and incorporate great heterogeneity, which fully justifies the deployment of secure Web services and moreover demonstrates their power in real-life applications.

4.3   P2P Networking

The approach discussed so far, primarily deals with a computing environment with entities interacting in a classical client-server model. This section provides a brief discussion of an alternative Peer-to-Peer (P2P) architecture [182]. Pure P2P networks merge the roles of client and server, rather than delegating responsibilities to a few centralized server machines. Hybrid forms of P2P and client-server architectures involve some central hosts, which store information about the peers registered in the environment. P2P environments are very well known from file sharing applications, such as the Gnutella network, and are typically useful for establishing ad-hoc connections. Advantages of P2P technology is resource sharing. Since many computing nodes are available, there is no longer the need for expensive and powerful centralized server machines. Adding more nodes to the network increases the overall system performance, while the opposite is true for client-server architectures, since an increasing number of client nodes reduces server throughput. Network management on the other hand is not necessarily a drawback, because large networks often require many servers, these nodes have to be coordinated as well. From a legal perspective, P2P networks always ranged in some grey area due to copyright issues mainly driven by the music industry.

If we wanted to apply P2P technology in SemCrypt application scenarios, some considerations need to be made. Access control policies and encryption keys have to be protected from unauthorized access at any circumstances as explained so far. Key management may be deployed in a still flexible manner though, since keys may be stored at the security server or on behalf of the clients. Considering hybrid forms of P2P networks, we still may deploy a security server, which takes care of authorization management and optionally of key management. Peers could take over the role of the insecure storage provider. If we assume that data involved in SemCrypt applications are secured in a way that there is no possibility that an unauthorized party can use them, it is save to swap the data on client machines, regardless of

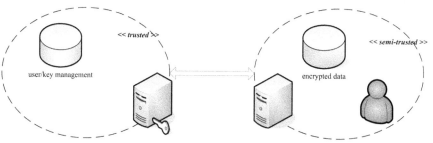

Figure 4.2: Hybrid P2P networking architecture

whether they are trusted or not. Such a hybrid scenario does not seriously affect application deployment so far, since instead of querying a centralized data store, requests are sent to other peers. Thus we resolve the triangle structure as illustrated at the beginning of this chapter by simply moving the untrustworthy data storage to the client side as illustrated in Figure 4.2.

More challenging are pure P2P networks, since we have to think about security management more carefully. Access control policies need to be protected at all costs, which means that we cannot store them on peer nodes without further measurements. We need to ensure that a subject never gains access to its own access rights it could modify. The only possibility of P2P settings is thus storing polices on other peers raising the question which access rights these policies may contain. A malicious peer could host and arbitrarily modify policies and thus inject permissions or denials, which would throw the application into inconsistency. A solution to that could be to couple access control policies with the actual data they refer to. If a peer stores only policies, which apply to the data it owns, it will not want to maliciously modify permissions, since they affect the peer's private data. We must therefore ensure that each peer only provides its own sensitive data and additionally corresponding authorization policies and cryptographic keys, which are eventually required for accessing the data.

However, there are many more security issues that arise in P2P networks and which cannot be solved by SemCrypt technology. P2P environments often suffer from poisoning and polluting attacks, which refer to peers providing data containing differ from their description or data containing bad chunks or even mal-ware or viruses making the whole file useless and even dangerous for the recipient. Other forms of attacks are denial of service, spamming and defection attacks, which must be managed separately.

Chapter 5

Analysis and Design

This chapter contains requirements analysis of security components, which are mandatory in SemCrypt applications, i.e. access control and encryption and decryption modules, followed by concrete design suggestions implementing the security functionalities. The next chapter, on the other hand, discusses how these security components can be invoked in a service oriented manner using Web service interfaces. All these responsibilities are organized within the SemCrypt Application Framework (SCAF) [140]. The SCAF serves as umbrella for security components and the service environment. Security components are namely the SemCrypt Authentication and Authorization Framework (SCAAF) [142], the SemCrypt Encryption and Decryption Framework (SCEDF) [144], which are both subject of the next two sections, and the SemCrypt Database Framework (SCDBF) [143], which represents the secure persistency layer for storing authorization policies, cryptographic keys and user and document information. The service environment is divided into the SemCrypt Web Services Framework (SCWSF) [148] and the SemCrypt Security Services Framework (SCSSF) [147], dealt with in Chapter 6.

5.1 Access Control

So far, related work has been discussed and put into context with the goals of this thesis. Another focus was on identifying application scenarios, which are suitable for secure XML data outsourcing, where HRM has been chosen as reference show case. The following chapter is maybe most important, since it deals with requirements for the concrete implementation of the access control component, which must be hosted by a trusted security server. Advanced XPath features are introduced to establish a context with the authorization procedures described further on. The access control language being developed in the course of this chapter uses a reduced set of XPath syntax to express which XML fragments must be protected. Reducing the XPath syntax facilitates policy decisions and enforcements, since it is easier to manage. On the other hand the access control mechanism must not be trivialized. A preferably large set of XPath queries should be allowed in order to not be forced to reduce the areas of deployment of the access control component and at the same time not limit the power and expressiveness of the XPath query language.

5.1.1 Authentication

The SemCrypt Authentication and Authorization Service (SCAAF) specifies components for authenticating subjects, willing to interact with the application, and authorizing their requests for accessing the untrustworthy database storage provider. Authentication forms the foundation for any further request processing and is thus fundamental in any application setting. Without identifying the invoking client party, no further processing is possible. Authorization is always performed directly after authentication to check whether the client has sufficient access rights on the database to perform the desired operation.

The SCAF reference implementation provides a custom authentication mechanism, which is based on the Java Authentication and Authorization Service JAAS [156], and thus provides all the benefits JAAS offers to the application developer. Since authorization is a fundamental subject to SemCrypt and JAAS is insufficient for providing access control on all levels necessary within the SCAF, this section is focused on the authentication process only. Figure 5.1 illustrates the most important components involved in the user authentication process: Login process starts as soon as an instance of the `AbstractUserInterface` implementation is invoked. For improved readability method listing has been omitted, but the process is rather straightforward, since it basically is restricted to calling operations provided by the JAAS classes. The `AbstractUserInterface` is responsible for collecting the user authentication data. In this case we focus on password based authentication and provide username and language code as well to the SCAF. The login module instantiates the authentication components and passes on its own reference to allow the `DefaultCallbackHandler` querying the login information. For the sake of completeness, JAAS authentication components are depicted as well. The `LoginContext` provided by the JAAS API is the glue between the `LoginModule` and the `CallbackHandler` omitted in the diagram.

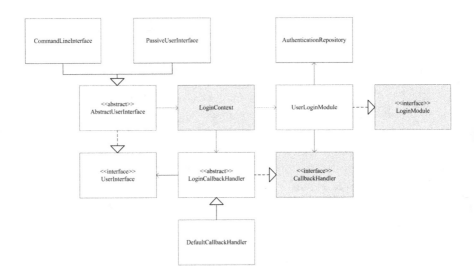

Figure 5.1: Authentication components in SemCrypt applications

Both are JAAS interfaces as well and need to be implemented by custom classes. To be more concrete the LoginContext expects a CallbackHandler and a configuration file as input. The configuration file in turn specifies which login modules to use for authentication.

Although JAAS allows the specification of a set of login modules, for now we use one single module named SCLogin. As specified by the JAAS API, there are several ways how to tell the LoginContext about the login modules. Since we know our configuration file, which we may wish to modify in the future, the most convenient way in this case is probably to do it programmatically by setting the system property java.security.auth.login.config to our configuration file path during Java Virtual Machine (JVM) initialization. Besides the login modules, the configuration may contain space separated additional parameters which may be accessed by the login module's initialization method. As soon as the login method is called on the LoginContext it is passed on to the login module, which creates all Callback objects necessary, such as username, password or language. The CallbackHandler then gathers the required information from the user and forwards it to the login module, which performs the final authentication.

Note that the AbstractUserInterface can be regarded as interceptor be-

tween the user interface and the actual authentication module. This enables
higher flexibility, since we only need to integrate the LoginInterface with
any client frontend and do not need to rewrite the callback handler for each
login method anew. Though callback handling using command line user
input is trivial, it would require much more workarounds for Web based
user authentication due to the request/response mechanism of Java Servlets.
Identity management is one main issue specified by the SCAAF. Authentication is needed for every incoming client message and must be performed
on the service layer before forwarding the request to the internal components, such as authorization and key management. Authorization is enforced
by means of an XPath statement, which aims at a specific portion of data
stored in the database and which has to be analyzed to determine whether
the requesting client is granted access. Different access control approaches
are discussed and compared regarding their applicability and extensibility in
SemCrypt applications.

5.1.2 Query Processing

XPath is a widely applied technology in the XML area. Many efforts have
been made to optimize XPath expressions to speed up query processing and
evaluation. Although the term is often put into context with the database
world, query processing is of significant importance in other fields, such as
access control, which shall be discussed in the following section. Query processing is performed using a two phase model, which involves static and
dynamic query evaluation.

Static context information is available during expression analysis before evaluation is actually performed. This is where query authorization is hooked
in, since whether a request is granted or denied should be known, if possible, before the query is forwarded to the XML store. However, the reason
why this has to be said with some caution is that sometimes post filtering of
query evaluation is required to be absolutely sure that no access policies are
violated. These issues will be subject to the next section. Static evaluation
comprises function names, namespaces, document, in-scope schema definitions, types, which are statically known. Dynamic evaluation covers function
and variable values, dynamically available documents, date and time values
and most importantly context processing, referring to the context element
currently being processed. XPath query processing is a very complex task
and has been subject to many scientific publications. For instance, Gottlob
et al. propose algorithms which run in main-memory with polynomial time
complexity and introduce XPath fragments for which linear-time processing

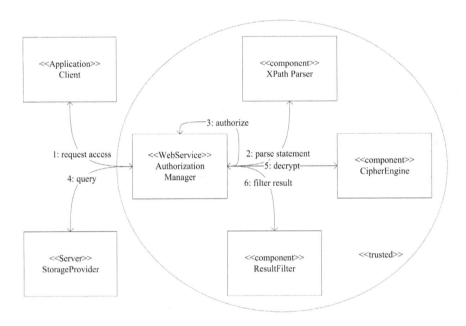

Figure 5.2: Query processing in SemCrypt applications

algorithms exist [75].

According to the information provided so far and the architectural requirements analysis from Chapter 4, the following paragraphs are dedicated to
authorization processing. The following image reveals the interaction among
software components necessary for XML based access control using a simplified collaboration diagram: Components are separated into trusted and untrustworthy domains, i.e. clients and database server. Trusted components
involved in query processing are the authorization manager, cipher engine,
query parser and result filter. The client is representative for any application
interacting with the trusted Web service, such as a standalone application or
a service provider enabling access via an intermediary application framework.
The client application, however, is responsible for generating appropriate request statements that can be handled by the Web service component, which
are XPath statements. The authorization manager forwards any incoming
request to the XPath parser which validates the statement and returns an
object hierarchy similar to the document object model [170]. The resulting
DOM is used to perform the authorization process. One big advantage of

the object model is reusability: each valid XPath statement can be compiled into objects, which highly facilitates working with XPath expressions on a programmatic level. Existing XPath parsers fail to provide these interface, e.g. [47]. Furthermore its modular design allows comfortable application integration. The drawback, of course, is decreasing performance, since requests have to be parsed twice: once to retrieve the object model and once again traversing the object hierarchy during authorization process. A compromise to increase performance without losing ease of use could be to just generate absolutely required objects.

After parsing the request, the result is corrected by eliminating context items, i.e. access to the current node within the expression context, reverse steps forward steps, axes, type casts, variables and function calls. In other words queries are transformed into valid canonical XPath expressions by extracting only absolute path expressions and eliminating axis navigation and abbreviated reverse steps. Function calls are considered as irrelevant as well for authorization purposes, since if any path relevant for a function call is accessible, there is no reason why the operation on the path expression should be prohibited. We actually could restrict on return values of a function and due to the openness of our proposal, this could be subject for further extension. Casts and variable assignments may fail during XPath evaluation but are also not subject to security considerations. he discussion of related work in Chapter 2 provides XPath rewriting examples, which serve as starting point for the design of the authorization engine.

Query transformation makes it much easier to check for access permissions, since irrelevant query information are ignored and only absolute path expressions and its contents are of main interest. Since each expression may recursively contain sub-expressions, i.e. predicates, a take bottom-up approach is taken to evaluate expressions on the lowest level first to be able to evaluate the parent expressions. For example, if we defined an access restriction on an employee's monthly wage, we first need to know what wage the predicate is pointing to before we can permit or deny the access. `//Salary[/SemCryptEmployeeInfo/PersonInfo/PersonId[IdValue = "9902261"]]` for instance, would never make it through the authorization manager if the requested wage is higher than the requestor is allowed to see. The biggest drawback of this approach is again performance. Very complex queries waste a large amount of network traffic, since a potentially vast number of sub-queries are being made to the untrustworthy database server. And in the worst case an access denial may occur at the final stage of evaluation, which implies error detection unnecessarily late. Nevertheless, the algorithm operates on exactly relevant data. Another possibility would be to forward all top-level path expressions to the database, which causes as many requests as the query contains top-level expressions. `//PersonInfo/PersonName` and `//PersonInfo/ContactMethod` would generate exactly two database requests. In the worst case, requesting XML root elements would result in transfer of the complete database content. The listing below illustrates the basic bottom-up authorization approach in pseudo-code:

Listing 5.1: Authorization algorithm for selective access

```
FUNCTION authorizeSelect expr
    FOR each pathExpr in expr
        IF pathExpr contains Predicate THEN
            INIT resultList
            FOR each Predicate in pathExpr
                CALL authorizeSelect with pathExpr RETURNING result
                ADD result to resultList
            END FOR
            CALL rewriteExpr with pathExpr and resultList
                RETURNING pathExpr
        END IF
        CALL evaluateExpr with pathExpr RETURNING granted
        IF granted EQUALS true THEN
            INVOKE DBServer with pathExpr returning cipher
            INVOKE CipherEngine with cipher RETURNING decrypted
            INVOKE ResultFilter with decrpyted RETURNING result
            RETURN result
        ELSE
            RAISE AuthorizationException
        END IF
    END FOR
END FUNCTION
```

The function takes a corrected XPath statement as input and calls itself as long as the algorithm discovers Predicates contained by the expression. If the lowest level has been reached, the path is evaluated by the authorization component and sent to the untrustworthy database server. If we wish to gain read access, the resulting data is encrypted and has to be decrypted by the cipher engine. For validation purposes we filter the resulting data by applying the XPath request again on the result and return it to the calling function. Formally speaking, a request of form $p_1[p_2 \text{ op } p_3]$ where p_i denotes a canonical path representation and op some comparison operation, after replacing the rightmost path p_3 or leftmost path p_2 with the appropriate evaluation result, makes it possible to evaluate access control to p_1 with respect to its restriction definitions. If we turn back to the collaboration diagram, the invocation of the cipher engine depends on the type of action requested. We can reduce database access requests to reading and modifying access, regardless of which access rights have been specified previously. Anyway, the cipher engine has to be called to encrypt data which is about to be inserted or to decrypt data that

XPath	:=	PathExpr
PathExpr	:=	("/" RelativePathExpr?) |
		("//" RelativePathExpr) |
		RelativePathExpr
RelativePathExpr	:=	StepExpr (("/" | "//") StepExpr)∗
StepExpr	:=	AxisStep | FilterExpr
AxisStep	:=	(ForwardStep | ReverseStep) PredicateList
ForwardStep	:=	(ForwardAxis NodeTest) | AbbrevForwardStep
AbbrevForwardStep	:=	"@"? NodeTest
ReverseStep	:=	(ReverseAxis NodeTest) | AbbrevReverseStep
AbbrevReverseStep	:=	".."
NodeTest	:=	KindTest | NameTest
NameTest	:=	QName | Wildcard
Wildcard	:=	"*" | NCName ":" "*" | "*" ":" NCName
FilterExpr	:=	PrimaryExpr PredicateList
PrimaryExpr	:=	Literal | ParenthesizedExpr | ContextItemExpr
Literal	:=	IntegerLiteral | DecimalLiteral |
		DoubleLiteral | StringLiteral

Table 5.1: The subset of XPath syntax used for access control description

is going to be displayed to the user on the client application frontend. Calling the result filter for validation purposes is optional but provides additional safety regarding sensitive result data. Modifying the algorithm shown above to suit the requested action should be straightforward.

5.1.3 XAccess Control Policies

Policies describe permissions and/or denials for XML document access. Both permissions and denials are made up of a triple <S, A, R>, where S denotes the subject, i.e. the role or user for instance, making the access attempt, A refers to the action the subject wants to perform, such as read or write, and R is short for the resource the subject is trying to access. Resources are specified as XML fragments, which may be selected using XPath. The table 5.1 shows the subset of the official XPath syntax, which is valid for describing the XML resource: The allowed syntax is very restricted, but sufficient to describe and protect all possible XML fragments. A standard conformant XPath expression can be made up of multiple path expressions. To perform authorization checks it is sufficient to investigate the smallest logical units of XPath expressions, i.e. step expressions, which make up a path referencing specific portions of XML data. It seems illegitimate to allow access to XML content and at the same time restrict the access of a sub-portion. For example, in case access to element E is be granted, while applying a substring function on E is prohibited, a subject may retrieve element E and apply substring autonomously, resulting in annoying workaround without providing additional security. Hence, XPath expressions are considered legal as long as the path expressions involved are allowed to be performed. Excluded from authorization processing are therefore, qualified, if, logical, arithmetic, union, intersection, instance of and cast expressions as well as functions. Filter expressions take up a special status, since they are disallowed in the XAccess path syntax, but need to be treated somehow. Filter expressions allow access to nodes via indices, node names or comparison expressions.

- /E1/E2[2] returns the second element E2 as defined by document order

- /E1/E2[E3] returns elements E2 with child E3 in document order

- /E1/E2[E3 eq V] same as above, but E3 must equal value V

The XPath specification allows whole XPath expressions to be used with filter predicates, i.e. considering the filter expression rule in the previous syntax table, a predicate list contains zero or more predicates, which are defined as [Expr], which in turn can evaluate to any valid XPath expression. However, reducing these XPath expressions to path expressions is again sufficient for making authorization decisions. Filter expressions provide additional granularity to permission definition, since not only authorization on element or attribute level can be made, but also restrictions on their contents may be defined.

Recalling the triple <S, A, R>, a resource R is thus made up of path expressions and restrictions, i.e. definition of filters. The reason why filters are excluded from the grammar specification above is that recursively defined path expressions makes permissions considerably hard to handle. An example should make that clear:
/E1[E2[E3 eq V]] is a valid XPath expression, but during evaluation time it is unclear what the contents of E2, E3 and possibly also V are. Such an expression has to be recursively resolved to decide whether the query points at legal XML fragments. Though legal with respect to the path expression grammar, using indices as filter is discouraged, since it cannot be known during runtime, how an element at a predefined position looks like, i.e. whether it holds content that needs to be protected. This is because of the semi-structured nature of XML, which describes the structure of documents, but not orderings of elements of the same name. Hence if subject S has access to

some element E with value V, a permission definition of E[2] may have fatal consequences, since there is no way to specify that E holding V must always be on second position regarding document order. This situation, however, can be prevented by not specifying permissions or denials for indices, but what if subject S queries for exactly the second element in document D? In this case, index based permissions could suddenly make sense to make access control decisions during static checking. Or the query may be allowed for now and the result is subject to post-filtering before being returned to the calling subject. Only performing static query checks, however, is only safe, by either prohibiting index-based queries or storing metadata about each document in the database, such that an index-based query can be mapped to actual element or attribute values.

The policy file format developed within this section uses XML to describe access control policies. XAccess allows the definition of permissions and denials, which both specify the same element structure. The difference lies in the policy processing semantics. Not surprisingly, permissions express paths that may be accessed and denial exactly the opposite. XAccess prescribes a restrictive handling of permissions, i.e. anything that is not explicitly permitted is prohibited. Denials may be used to overwrite permissions. For instance subject S may have access to element E1 but not element E1/E2. Since permission for a node N involves access to N and any sub-nodes, denials can explicitly restrict permission definitions. This can become especially useful in case of permission inheritance. To facilitate policy management, subjects with similar permissions may inherit them from each other. With denials it is then possible to restrict specific permissions on a very fine granular level, instead of requiring the whole permission to be redefined for another subject with only a few modifications.

Listing 5.2: Permission format of an XAccess policy file

```
<Permission>
    <PathValue />1
    <SubjectName />+
    <RightName />+
    <RequiresSubjects>
        <SubjectName />+
    </RequiresSubjects>?
    <Filter>
        (<Index />1) |
        (<PathValue />1 (<Operation />1 <Value />1)?)
    </Filter>*
</Permission>+
```

The previous code sample illustrates the info-set of an XAccess policy. Path values contain a string value as defined by the reduced XPath grammar at the beginning of this section. For each permission at least on subject must

be specified to which the permission applies. Also at least one access right
must be granted. Interpretation of the access rights is up to the authorization component handling the policy files. Required subjects aim at subjects,
which must be additionally provide their identity in order to allow the request, stated by some privileged subject, being processed. Filters are exactly
what have been discussed previously. For each path an arbitrary number of
filters may be specified containing wither a node index or a path value. If
an additional operation is present, a value is mandatory as well for defining value barriers of the path specified. Denials may be declared exactly
the same, whereby the processing sequence of the underlying authorization
engine should assign denials a higher priority than permissions. In case
of contradictory denials and permissions definitions, the denial should gain
higher importance and override the permission. Both permissions and denials are surrounded by a Permissions or Denials tag, which at least contain
a permission or denial respectively. Permissions and denials are contained
by the XAccess root element.

Another element which is subject to XAccess policies are namespaces, defined
as follows:

Listing 5.3: Namespace definitions in XAccess policies

```
<Namespaces>
    <Namespace>
        <URI>anyURI</URI>1
        <Prefix>string</Prefix>?
    </Namespace>+
</Namespaces>1
```

What has been omitted from the simplified XPath grammar table are namespace prefixes. These are defined by grammar elements not contained in this
table, such as the literal constants and qualified names (QName). A path
expression consists of step expression, whereby a step expression often refers
to an element name. Since XML elements may be defined by namespaces
other that the default target namespace, XPath may also contain namespace
prefixed path steps. Namespace information is extracted for permission definition, since the authorization engine must somehow be able to determine,
which namespace an element belongs to, in order to distinguish between eventually conflicting element names. Elements may be defined with the same
name but different namespaces. Hence, subject S may have access to element
E of namespace NS_i but not E of NS_j. Additionally, XML documents that
in fact are valid for the schema description, do not necessarily need to use
the same namespace prefix that is used in the schema definition. Namespace declarations are tuples <P, U>, where P denotes the namespace prefix
and U the namespace URI. For a document D valid according to a schema

S, namespaces are perfectly equivalent if URI(D) equals URI(S). For that reason a mechanism for mapping namespaces during request processing is mandatory. XAccess requires at least the definition of a target namespace.

5.1.4 Performing Authorization

Authorizing XPath statements is one central security issue in SemCrypt environments. Since XPath is used to select fragments of XML document data, which might contain confidential information, it is necessary to analyze which data a query is about to select and permit or deny the statement accordingly. In case of an XUpdate request, the containing XPath statement is first extracted and then passed on to the parser. Each request must contain a query context object provided by the API, which in turn contains at least one role actually stating the request. This is mandatory, since some requests may require multiple privileged users and simply knowing of currently logged in users is definitely insufficient for performing authorization. Starting point for defining access control policies is an XML schema, declaring the structure of the affected XML documents access should be restricted to. With schema information it is possible to determine the corresponding valid document structures and derive all possible element paths valid for a given schema.

This section deals with policy enforcement concepts in SemCrypt applications. Since the reference implementation is written in Java, authorization management is also based on the JAAS framework as introduced in the previous section. Core classes provided by the JAAS API are again illustrated as grey filled rectangles, while SemCrypt extensions are kept transparent. Entry point for authorization processing is again the UserInterface interface, which additionally to login and logout methods provides an abstract perform operation, which attempts to perform some application specific action by first checking user privileges with existing policy files through the JAAS layer. Additionally to the LoginContext, which has already been discussed previously, JAAS contains a set of abstract classes, which need to be extended to meet application specific requirements. The XAccessPolicy class is responsible for loading and parsing XAccess policy files and providing their content to the authorization framework. The idea of the CompositePolicy is to hold a collection of different types of policy files and not restrict the SCAAF to processing only XAccess policies. Currently, default JAAS policies and XAccess policies are supported. After parsing the policy file, an XAccessPolicy holds a set of XPathPermissions. Whenever an incoming user request involves XPath authorization, an XPathAction is created and checked with the permissions obtained via the XAccess poli-

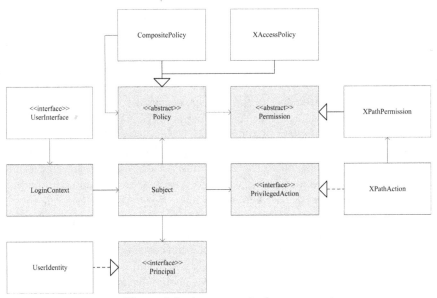

Figure 5.3: Access control components

cies. The whole policy enforcement process is done via the Subject object of the JAAS framework. After successful login via the LoginContext, the authenticated subject may be obtained, which may be associated with a set of identities. Identities are represented by the Principal interface, which is implemented by the SCAAF UserIdentity class.

The algorithm presented by Miklau and Suciu [112] is very promising also for access control processing, since it is capable of handling descendant selections, wildcards and filter expressions. With those features supported, it is possible to compare a wide range of XPath queries with respect to access control, since as discussed in previous sections it is sufficient to concentrate on path expressions anyway. With the previous sections in mind it is possible to overcome these shortcomings to a certain extent. In the following extension mechanisms to the algorithm by Miklau and Suciu shall be discussed. Axes can be handled by query rewriting and replacing axis navigation syntax with abbreviated syntax. Since path permissions as specified in the access control policies contain only abolsolute path values, even abbreviated parent (" . . ") and context navigation (" . ") can be eliminated by rewriting /E/../E to /E and /E/./E to /E/E. The problem remaining and also omitted when discussing axes and abbreviated syntax equivalence, are the preceding axis, which involves the preceding-sibling axis and the following axis, which involves the following-sibling axis, having no abbreviated correspondence. This problem is similar to the indexing problem presented earlier. Due to the semi-structured nature of XML it is a priori not possible to determine what elements are being accessed. Also here, the solution would be wither to entirely deny queries containing such syntax or perform post-filtering or store metadata about each document that could possibly be accessed. Metadata information must on all accounts contain data about the elements and their contents to make access control decisions during XPath processing. Post processing during XPath evaluation phase maybe would be much simpler to achieve, but requires an additional processing step by the authorization engine, which could result in unnecessary data transmission and performance overhead. However, the approach followed within this thesis is an extension of the algorithm by Miklau and Suciu with extensions regarding query rewriting and adding comparison expressions. The algorithm is capable of handling filter expressions on arbitrary levels of recursion, which is not required herein, since permission definition aims at fine granular XML access control, where recursive predicates are considered redundant and moreover hard to handle and understand.

5.2 Cryptographic Requirements

The SemCrypt Encryption and Decryption Framework (SCEDF) is responsible for encrypting and decrypting database content. Both symmetric and asymmetric keys may be used with the SCEDF. With asymmetric keys it is the application's responsibility that the public key is used for encryption and the private key for decryption. Inspired by preceding work, such as [113], the following sections are dedicated to the discussion of suggested encryption/decryption procedures as well as key management, which are subject to testing and evaluation within the SemCrypt framework.

5.2.1 Encryption and Decryption

Cryptography is one of the core components in SemCrypt applications. They are decisive with respect to the overall application security. It is not the aim of the SCEDF, to define innovative encryption algorithms or provide some formal cryptanalysis. It rather relies on existing cryptographic algorithms for symmetric and asymmetric key generation as well as their appliance in application management. Furthermore, this section does not describe the usage of specific algorithms, but rather a framework which is open for adding any security algorithm. Security with respect to data confidentiality thus depends on the strength of the encryption algorithms used for a specific application. It is the responsibility of the application developers and end users to choose algorithms which are secure and fast enough with respect to the actual application requirements.

The SCDBF describes the feature allowing each user to upload own documents and securing them with own cryptographic keys, which may either be stored at the security server or locally at the client side. Each user must therefore also have the possibility to upload own cryptographic keys, which may then be used by the security server to encrypt/decrypt document data and return the result to the client front-end. How key management is incorporated into SemCrypt applications is part of the service environment described in the next chapter. Key management responsibilities are delegated to subjects in charge, i.e. system administrators for general purpose ciphering, roles and single users. If a key is removed, appropriate actions have to be performed on affected artefacts, i.e. XCipher descriptions discussed in the next section and document data. Any references to the removed key have to be resolved by deleting them from the XCipher instructions and decrypting affected data fragments. If a key gets lost it becomes very hard if not impossible to recover the original data later on. These actions need only be taken, if the affected key is stored at the security server, otherwise it can be assumed that it is stored somewhere else and it is up to the client to know what to do. Dynamic key change, i.e. updating or modifying a key during runtime has the same implications as when a key is being removed and a new one inserted.

Since data is stored encrypted in untrustworthy database domains, the keys for encrypting/decrypting the data as well as required cryptographic procedures are locked away on the security server. Access is only granted if the requesting party could be successfully authenticated. Key management and ciphering techniques are part of the SCEDF [144]. It discusses how data confidentiality is ensured within the SCAF and falls back on some methods developed by the SCAAF, such as XPath processing, which is already required by encrypting/decrypting sensitive data.

Appendix A contains a schematic listing of the XCipher language. It specifies an XML descriptor containing instructions of which key should be used to encrypt a specific portion of XML data. The descriptor is very simple. The root is the XCipher element, which contains at least one `Encryption` element. Each encryption element in turn contains at least one path value and at least one key alias. The key alias must be a string value identifying exactly one cryptographic key in the client or security server key store discussed in prior sections. The path value is an XPath compatible expression (see Chapter 2) of the form:

```
XP := ("/"(N":")?E)* ("/"(N":")?  "@"A)?
```

If multiple key aliases are specified, the identified data portion is being encrypted exactly in the sequence the keys are specified. When decrypting the affected data, the inverse sequence is taken. Any modifications to the XCipher description during runtime without proper configuration may cause unstable application states, which eventually makes data irrecoverable. If multiple path values are specified the key sequence must be applied and the ciphering algorithm works as follows:

1. Apply the XPath containment and equivalence algorithm discussed in [112] and order the path values by their containments.

2. Obtain the referenced cryptographic key from the security key store if possible.

    (a) If keys could be retrieved, apply the encryption algorithms bottom-up (top-down in the decryption process) on the ordered path values.

    (b) If keys could not be retrieved, indicate an error and return the unprocessed plain text (cipher text in the decryption process).

3. Repeat steps 1 and 2 for each encryption element contained by the descriptor.

A note on step 2: The encryption/decryption processes are defined only if the relevant cryptographic keys are stored either at the security server or at the client side. In the latter case the XCipher descriptor is more or less a reminder for the client to lookup the ciphering instructions, but actual encryption/decryption must be completely performed at the client side. Although imaginable it is not yet part of the SCEDF specification to split up the ciphering process being performed mixed at both the client and the server side.

An encryption element can thus be used to define an encryption/decryption sequence. Multiple path values may be added on which the key sequence should be applied. Since it may happen that a path defined is contained by another path it must be ensured, that the sequence is applied on the path denoting a subset of another path first. Otherwise the node-set denoted by the parent path is encrypted and the child path can no longer be accessed. The same procedure has to be followed with the other encryption elements.

5.2.2 XCipher Instructions

In the last section we mentioned the trusted cipher engine, which is being dealt with greater detail in this part. The untrustworthy database management system has been designed to support multiple cryptographic keys, allowing XML elements encryption differently and even multiple times. As described earlier in this thesis, design executives who are planning and deploying an application scenario may specify very fine granular access control for XML documents. Key management is meant to be part of the design process as well. For each path that is valid for scenario documents, the application designer may define at least one key the specified path is being encrypted with. During runtime, the trusted domain is able to determine, which element is affected by the request and which key to choose to encrypt or decrypt the data. Since we already authorized the request in a previous step, we just apply the correct key to the data without needing to associate the requesting role with the cipher key. We store key information in a separate XML file, called `xcipher.xml` on the trusted domain. The most relevant information is listed as follows:

Listing 5.4: The XCipher encryption instruction info-set

```
<XCipher>
    <Encryption>
        <PathValue>string</PathValue>+
        <Sequence>
            <KeyId>string</KeyId>+
        </Sequence>1
    </Encryption>+
</XCipher>1
```

Since multiple encryption of XML elements and multiple keys per schema are allowed, a ciphering instruction is defined as a collection of at least one path valid for a schema and an ordered list of cryptographic keys referenced by a unique identifier. For each path p_{i_j} in encryption instruction ks_i, keys k_k are applied in the order they are specified in the descriptor. Manipulation of this description during runtime without performing dynamic key change on the document provider side, would make encrypted data become useless since not restorable by the application logic. Key sequences are handled sequentially as well. Similar to the user database, keys are also stored in a relational database at the trusted domain and may be identified via a unique key identifier formatted as string.

Key storage is assumed by a local key store. Each key is identified by a unique alias, an array of binary data actually containing the key data, an optional password for protecting the key from unauthorized access and a list of certificates, which is only required for protecting private keys in case of asymmetric encryptions. Implementation of a key store is application responsibility, however, the default SCEDF implementation incorporates a Java Keystore [157], which exactly states and implements these requirements.

5.3 Security Data Management

Security information needs to be available throughout the whole application lifecycle, which forces the call for data persistency on the security server. This is specified by the SemCrypt Database Framework (SCDBF). Data storage involves user information, key management and document metadata using relational data tables as well as XML formatted policy files ciphering and versioning information held by a native XML database. The SCDBF describes the SemCrypt persistency API with special focus on its flexibility regarding the deployment of concrete database back-ends. This includes the relational database scheme describing entity attributes and relationships as well as XML document organization.

5.3.1 Storage Abstraction

The purpose of the SCDBF is to provide a programming interface for both relational and document centric databases. The database framework serves as umbrella for any underlying database technology that might be deployed. Each type of database has a corresponding database manager attached, which is responsible for creating, deleting as well as opening and closing a database connection. Semantics of those operations differ depending on the concrete database implementation. Diagram 5.4 illustrates the interrelations of database components using UML class diagram notation using the Abstract Factory Pattern [73]. The basic database declaration does neither contain any operations or global fields, since the technological nature of relational databases is fundamentally different from document oriented XML databases. Methods provided for relational databases comprise CRUD implementations for all types of database tables, while XML databases incorporate document management. Although the SCDBF provides a uniform way for handling different types of databases, it is not yet possible to provide operations independent of the underlying technology. The operations defined for XML databases are completely different from relational database

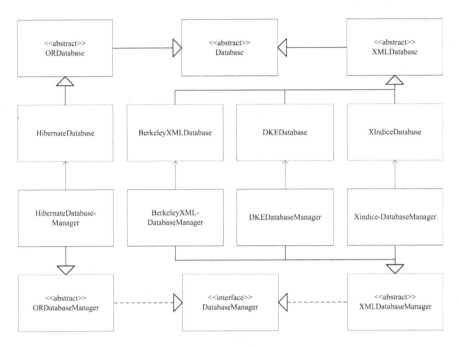

Figure 5.4: The SemCrypt database components architecture

operations. The relational database layer serves as wrapper for underlying object-relational mapping technology, such as Hibernate [131], TopLink [127] or Cayenne [12], and does not directly operate with SQL statements, which tremendously facilitates development efforts. For the sake of flexibility and openness, the relational layer provides a well-defined interface for plugging other mapping frameworks. Object-relational mappers can push productivity. Unfortunately, at time of development no comparable technology for XML databases existed, which forced the implementation of different XML database handlers from scratch. Thus the application developer at least has to decide which type of database to deploy, the rest can be determined dynamically. The SCDBF currently supports XIndice [8], Berkeley DBXML [126] and an interface to the database storage provided by the DKE Linz.

The reason why the DKE interface has been incorporated with the XML database layer might appear some kind of weird, since SemCrypt data are not stored as plain XML, but rather as encrypted key/value pairs. The point is that the database layer hides the physical structure of the stored data and is focused on functional services. Neither programmers nor application users should be confronted with how the data is stored, what counts is the type of data and in the case of SemCrypt it is XML document data, which should be presented to the users. Hence the SCDBF provides a document centric interface for actual data storage, independent of the underlying XML database.

5.3.2 Hierarchical Document Storage

The document centric database interface is not only used for handling the untrustworthy document data store, but also deployed for holding application specific access control policies, ciphering instructions and versioning information. These files contain very sensitive application specific security information and must thus be stored at the security server. Since those artifacts are represented as XML documents, it appears reasonable to implement an XML layer for storing these files to profit from the benefits of a native XML store, such as document indexing and thus improved querying and updating performance. A possible XML storage structure is illustrated

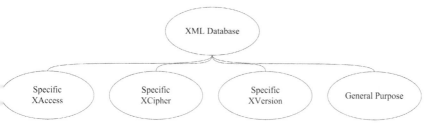

Figure 5.5: Suggested XML storage structure

in the tree graph 5.5 above. The top ellipse represents the application specific data store holding four document collections. Collections are an XML database mechanism to group related documents within a database. One collection could be used for holding the general purpose access control policy, ciphering instruction and versioning document, while the other collections represent specific storage containers. General purpose policies and encryption instructions apply to every document involved in the actual application setting. These general purpose files may be overridden by document specific artifacts, stored in the designated XML container. Files stored within the trusted XML database can be accessed via a unique identifier, which must be known to the relational database to reference the artifacts accordingly. For instance, if a document is secured by a user defined access control policy, the

policy is stored in the specific XAccess policy collection and referenced by
the document via its Id. However, the structure presented not mandatory,
but represents the default implementation and should clarify the purpose of
the data store. The application must only ensure that identifiers used for the
XML descriptors is identical with the identifiers used in the document table
of the relational layer, such that a document can be easily associated with
the corresponding XML metadata.

5.3.3   Versioning

Storing versioning information about XML document data was also considered in initial SemCrypt objectives shall be briefly discussed as secondary
achievement within this thesis. The approach taken focuses on the integration of existing Diff-Tools tailored to XML formats, such as the xmldiff
project presented in [50]. As mentioned in the related work in Chapter 2,
there are many research papers, which aim at research documentation in the
area of XML versioning, such as [41], which suggests a framework for assign-
ing different access control policies with different versions of the same XML
document. Although, this approach definitely deserves further investigation,
it has not been integrated in SemCrypt and would probably fill additional
Ph.D. theses. However, versioning has been considered regarding the research
agenda and may optionally be implemented in SemCrypt applications. An
XML schema for storing versioning information, called XVersion (to fit in the
naming conventions of the framework developed) has been suggested, which
is listed in Appendix A.

5.3.4   Relational Information Storage

The relational layer is not supposed to store sensitive document data but
provide persistency for application management data, i.e. store users, roles,
documents metadata and access rights. Figure 5.6 below illustrates database
entity relationships: How these entities are actually described depends on
specific application requirements. Nevertheless each entity must define a set
of attributes, which is listed in the following table.

- **User**: Id and credentials

- **Role**: Id and description

- **Document**: Id, according schema, XAccess, XCipher and XVersion

- **Right**: Id and description

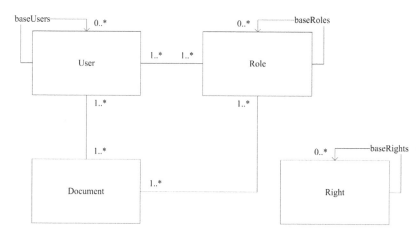

Figure 5.6: Relational entity relationships

Each application must be able to identify the users registered in the system. Hence, each user must be given the possibility to choose a unique username or the application uses an Id, which is per definition unique, such as a user's mailing address. For increased performance, however it is suggested to additionally use an internal identifier as primary key, which is represented by an auto-incremented numerical value and which is not visible outside the database layer. This also avoids database reengineering in case of merging several data stores and possible primary key conflicts when only using user-assigned Ids. In such operative decisions, uniqueness can no longer be guaranteed. Additionally to usernames, it is required to store credentials in order to perform authentication if a user logs into the system. For credentials a binary data type should be used to stay independent of credential types, such as encrypted passwords or private keys for instance. Users must belong to at least on group of users, called role. A role consists of an identifier and a textual description. For the role Id the same rules apply as for user Ids. The Id listed in the table above is meant as unique textual name of the role. Analogous to the user Id it is encouraged to use an additional internal numeric Id. With each role at least one user is associated, indicated by the n:m relationship between users and roles to facilitate navigation between these entities.

Each document is identified by a unique textual name and optional schema, XAccess, XCipher and XVersion fields. Typically a SemCrypt application supports documents, which validate against a specific XML schema. The database structure does not impose any restrictions, but is kept open to eventually support many different schemas. The name of the corresponding schema may be added to a document entry. For each application a set of permissions and/or denials are defined using the XAccess policy structure. One single policy file is sufficient to represent the general purpose permissions for an application. However, each user and each role may upload documents individually and may wish to define very specific access control policies for their documents overriding the general purpose access control. This introduces some notion of Discretionary Access Control (DAC) allowing the owner of a file defining permissions for other subjects involved in the system. The same applies to XCipher encryption/decryption instructions and versioning. When uploading specific policies or encryption instructions, these artifacts have to be associated with an already existing document. The XAccess, XCipher and XVersion fields are then modified to hold a reference to the specific descriptors saved in the XML data store. The order in which access control, ciphering and versioning rules are applied is basically the decision of the application, the initial assumption, however, is to override general purpose definitions with fine granular rules. Hence, if a user or role specific policy is found, it is taken, otherwise the general purpose definitions are applied to the document. To clarify the dependencies among the relational and XML data stores, the following figure 5.7 illustrates the XML store on the left hand side and the relational database on the right side. Both are associated via the document table, which optionally contains references to user or role specific XAccess, XCipher and XVersion documents persisted in the trusted XML storage. Managing these data and ensuring consistency and integrity is a very complex task and will be thoroughly discussed with the interfaces specifications later on. The database schema enables users and roles individually storing a set of private documents. User or role documents respectively are stored in separate associative tables as indicated by the n:m relationships.

The rights entity is optional in the SCDBF database schema, since it simply lists all relevant access rights for a specific application independent of the other database entities. Subjects, i.e. users and roles, and actions or rights, are brought together on a higher level in the application structure, i.e. on the XAccess layer which defines permissions as tuples of subjects, actions and resources, which are data portions denoted by XPath statements. Rights as well as roles have recursive relationships. This allows composition of existing roles and rights at increasingly high levels of abstraction. It enables more complex rights definitions, such as execute or delegate, on the basis of existing rights. A big advantage of role extension is inheritance of existing permis-

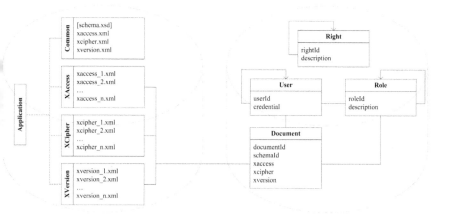

Figure 5.7: Trusted relational and XML storage interdependencies

sions and denials. This increases manageability of the database system. If after application deployment during runtime an additional group of users is added, which should be given the same permissions as an existing group with certain exceptions, it is not required to define the same set of permissions a second time. Hence we formally define a role as tuple R<id, R*> containing a unique identifier and a set of an arbitrary number of base roles. This may also be the case with newly added users. Furthermore it would be imaginable to introduce the same extension mechanisms to documents as well, with the confinement of allowing only one single base document, because of possible conflicts with access control policies and encryption/decryption instructions. For now this is not considered as part of the SCDBF specification but subject to further discussions and future extensions.

Part III

Implementation

Chapter 6

Service Environment

While Chapter 5 contained a detailed analysis of required security components, this chapter discusses appropriate service interfaces, which provide the functionalities to the outside world. Services discussed offer well-defined functionalities through state-of-the-art Web service interfaces. These services could be easily integrated in existing software requiring trusted persistency. Additionally, the services are not tied to special database systems, which facilitates integration in existing business workflows. The big advantage of XML orientation over traditional middleware is very evident in these cases. Independent of how the client application is actually implemented, it will work with the SemCrypt service environment, reaching from standalone rich client programs over Web applications to workflow integration. This chapter starts with general comments on the service environment regarding configuration possiblities and basic information about operations provided by the services. The SemCrypt Web Services Framework (SCWSF) [148] comprises services for the SemCrypt security components, i.e. access control and cryptography, and a service definition for directly accessing the unsecured data storage domain. The SemCrypt Security Services Framework (SCSSF) [147] implements very particular data storage requirements. It makes basic SemCrypt functionalities, i.e. XPath authorization and storage unit encryption/decryption operations available through a Web service interface as well as an RMI server and TCP sockets to speed up distributed database access.

6.1 Introductory Comments

Security services comprise authorization and ciphering components. The SemCrypt Web Serivces Framework (SCWSF) [148] describes how these functionalities can be made publicly accessible via Web service interfaces. It

discusses relevant operations provided to invoking client parties as well as
security issues regarding secure interaction, especially features such as message encryption, digital signatures and most importantly communication of
authentication information to identify clients and their permissions within
the service environment. The SCWSF concentrates on what should be made
publicly available, not in which way. Interacting parties are thus free to access the services in which application context ever, depending on the concrete
application workflow structure. The SemCrypt Security Service Framework
(SCSSF) [147] on the other hand, is specifically tailored to application interaction as originally developed in the SemCrypt project proposal to meet the
requirements of the database management system as discussed in [93].

The SemCrypt Application Framework (SCAF) [140] consists of several separately configurable and deployable application modules, which provide both a
programming interface for programmatic application integration and a well-
defined interaction interface allowing access to the modules in a preconfigured
manner. These interaction interfaces are described as Web services, i.e. each
module provides a WSDL description of the functionalities it provides. The
purpose of this chapter is to provide an in-depth description of each WSDL
interface and general requirements of the SCWSF.

Each service is discussed listing and specifying signature as well as functional purposes and behaviors of the operations provided. The Appendix, on
the other hand, contains a complete listing of the service WSDL definitions.
Types used are either defined by the XML schema namespace referencing
standardized XML Schema types, or the concrete WSDL namespace, denoted as target namespace. The following table lists all relevant namespaces
with according prefixes used throughout this chapter: Some operations do

Prefix	Namespace URI
xsd	http://www.w3.org/2001/XMLSchema
tns targetNamespace	http://semcrypt.ec3.at/services/types

Table 6.1: Namespaces used for the SemCrypt Web Service Framework

not require to provide an output value by default Since WS-I [175] prescribes output values as mandatory, for each of those methods a return type
tns:CommonResponse has been defined in the appropriate namespace to allow returning any arbitrary, application specific value. However, responses
are not expected by those operations and handling of these values are not
subject of the specifications. Each operation defines a service specific fault to

indicate any exceptional behaviour to the invoking entities. These exception types are defined as complex schema types and are subject to incorporation of the WS-BaseFaults [101] specification in order to provide a uniform error representation in future implementations.

In order to invoke the services specified herein, it is required to provide authentication information to most operations except the general purpose encryption/decryption operations. Security must be handled implicitly by these services using WS-Security token profiles as provided in Chapter 2. Which (if any) of these tokens must be provided, depends on the concrete deployment setting. WS-Security defines different profiles reaching from simple username/password authentication to more sophisticated SAML single sign-on and identity management. Thus authentication must be handled by the underlying application service, not on the application level of the concrete service implementations. However, each operation allows passing on a string-formatted security token for any application-layer purposes whatsoever.

While the previous chapters were dedicated to implementation details of security and storage modules, as well as interface definitions for accessing these components, the next chapters deal with the interaction of these services. The purpose of these so-called management services is to provide a simple interface for facilitating application development based on the core components. This is achieved by abstracting the details of component interaction and letting preconfigured services handling the interaction. The first service is called SemCrypt service, named after the underlying technology exactly representing its capabilities. For application development, however, there needs to be a way to set up the computing environment, i.e. users and roles involved in the application, document and cryptographic key management. This is finally the job of the Application service, which enables access to the trusted data store in a service oriented manner.

6.2 Component Interfaces

6.2.1 Authorization Service

The purpose of the authorization service is to provide access to the underlying access control modules required in SemCrypt applications. Note that access control is discussed in greater detail in and is not subject of the current specification. Briefly stated, we need a way to determine whether a user may be granted access to a portion of data stored in the untrustworthy

data store based on an XPath request and a desired operation, such as basic
database-related CRUD access method. The authorization service only needs
to provide a single method with two parameters (path value and operation),
which returns an access confirmation token on success or raises an exception
if the user is not entitled to perform the desired operation. This behaviour
is subject to each of the services involved in the service environment and is
further discussed in the previous chapter. Note that no user information,
which could be used for authentication is provided within the service operations. This information is passed implicitly to the service using WS-Security
tokens in the SOAP header. Available options are also further discussed in
the preliminary chapter. The authorization operation is specified as followed
and provided by the Appendix:

authorize
Currently, the authorization service only offers a single operation, which performs the authorization of a client request, which is formatted as string.
SemCrypt applications implement the authorization service in a way that it
is capable of authorizing XPath requests. Any client party invoking the service needs first to be authenticated in order to determine its identity. With
the subject identity and two parameters, i.e. request value and desired operation, it is possible to perform authorization checks. To stay SemCrypt
conformant, the operation is supposed to obtain XAccess policy descriptions
from the trusted data store and determine permissions and denials for the
invoking party as explained in the previous section. As already introduced
with the storage service, the operation also includes a string parameter, which
serves as placeholder for any application specific purposes, such as application specific authentication.

The operation accepts three parameters, where the first one is a placeholder
for any application level token handling, such as in case a user shall be
authenticated by a WS-Security independent authentication module. This
parameter is optional and is not handled in the current authorization implementation. The operation does not specify any return value, but indicates success or failure of the authorization process via a message indicating,
whether an exceptional state has occurred. The `AuthorizationException`
is specified within the schema part of the WSDL description.

6.2.2   The Encryption/Decryption Service

The encryption/decryption service has been designed as gateway to the underlying SCEDF [144], which implements the key management and crypto-

graphic procedures to encrypt/decrypt portions of binary data. These data fragments may represent XML document structures or binary storage units and ids respectively. The functionality provided by the cipher service may be accessed using the operations, which signatures are specified as follows:

encrypt

Encryption is usually done directly after the authorization of an XPath statement, whenever a user attempts to modify the database content. Authentication of the encryption operation should rely on ensuring that authorization process has already taken place or take care of authorization itself. The later option an be chosen, since the XPath expression must be passed on as string parameter to decide which key has to be chosen for encrypting the XML data, which is also delivered in string format. Since actual data upload is performed by some storage service, the encryption operation only encrypts the text and returns the result as byte array. Note that invocation of the cipher service is optional in cases the client overtakes the responsibility of encryption and decryption processes. Encryption fails and throws an exception of type `CipherException` in case no appropriate key is found on the security server to encrypt the plain text, encryption failed for some technical reason, or the user has insufficient privileges to access the service operation.

decrypt

Decryption of cipher text in SemCrypt application, follows the same principles than the encryption procedure. The XPath statement is required to identify the corresponding cryptographic key, and the cipher text has to be provided to apply the key. Calling the operation within an application workflow is optional if the client holds the required key. An error message using `CipherException` is returned to the client, if the invoking party has insufficient access rights, decryption failed due to technical errors, or no appropriate decryption mechanism could be applied to the cipher text.

Further operations defined by the cipher service are optional and only serve as general purpose methods for encrypting/decrypting plain- or cipher text. The user may transmit data to process and required additional information, i.e. most importantly key and algorithm to the cipher service and obtain the result formatted as array of bytes. Implementation of these operations is not part of the SemCrypt encryption/decryption framework and may be omitted.

6.2.3   Untrustworthy Storage Service

The service is intended to be used with databases other than the data store provided by the DKE Linz, since implementations should directly operate on XML databases.

Both, relational database management system and XML database are part of the trusted data and metadata storage layer and as such subject of the SCDBF. For accessing the untrustworthy storage server, which actually contains the probably encrypted XML document data, the SCAF also comes with a service interface, which allows direct interaction with the actual storage server. Although the deployment of this service is considered optional, it abstracts the integration of any kind of external data storage. And though not related with the trusted database layer the storage service is actually considered being part of the SemCrypt persistency layer and thus also subject of the SCDBF. It allows CRUD operations on XML documents and enables users to query and update the database content. Modifications of schemas contained documents adhere is not allowed, since such an action involves additional application logic that has to be performed in order to prevent the database from falling into inconsistent states. All storage service functions again include an optional application token, a mandatory document Id, which identifies a specific document in the database and update information depending on the desired operations. Appendix B lists the complete WSDL document of the storage service specified below. The storage service is supposed to serve as gateway between the untrustworthy data storage and invoking parties. It neither performs any authorization checks for query or update requests, nor does it care about encryption or decryption of document content. However, this service is part of the SemCrypt service environment and if deployed supposed to be properly integrated within the application workflow. Application developers need to take care about securing the service from unauthorized access very rigorously, since it does not prescribe the implementation of any additional security mechanisms. Like all services contained by the SCAF, however, also the storage service is supposed to be secured by an authentication mechanism, which must be configured in a way that only parties, which have properly accessed the authorization and cipher service before, may perform an operation provided by the storage service. Again, this can be achieved by appropriate WS-Security authentication mechanisms, such as SAML tokens or proprietary username/password identification. How this is actually done is the responsibility of the application developer, but it is strongly suggested to provide a proper trust domain for the storage service. The storage service provides a list of low level database operation, which shall be specified in the following paragraphs:

insertDocument
To add a document to the external, untrustworthy document storage, only the document's XML content formatted as string and a unique Id are required. Implementing components should only need to call the persistence mechanism on the underlying database management system and not care about document validation and similar issues. Most XML databases content themselves with document content and identifier. Database specific parameters required for persisting the document, such as transaction or context objects must be provided by the implementation, since the service shall abstract database specific features from invoking instances. The operation may return a `StorageException` message if a document with the Id provided already exists, accessing the database failed, or insufficient access rights have been passed on to the service.

updateDocument
Updating a document aims at replacing the content of an existing document. The Document identifier can thereby not be modified. Similar to the `insertDocument` operation, the update procedure abstracts database specifics from the outside world. A `StorageException` fault is raised, if no document with the Id provided exists, the invoking party could not be authenticated or internal database errors occurred.

deleteDocument
In case a document shall be permanently removed from the database, the caller of the service only needs to pass on the document Id, removing both identifier and document content from the database. An error message formatted as `StorageException` is thrown whenever the invoking party could not be identified, no document with the Id provided exists or for some technical failure occured.

selectDocument
Selecting a document means returning the document content as string. Once more, the implementing module must not implement any additional logic. Since underlying database APIs cannot be assumed to return the document content formatted as string, the operation implementation must take care of appropriate mapping mechanisms. A `StorageException` fault must be returned, if the requesting party could not be authenticated, no document with the Id provided exists, or the underlying database management system could not be properly accessed for some technical reason.

query

An XML query selects a portion of XML from the document specified by its unique identifier. Both, Id and query statement must be formatted as string. The actual syntax of the query depends on the query language of the underlying database management system. For instance, Apache XIndice [8] uses XPath 1.0 while Berkeley DBXML [126] supports XQuery and thus XPath 2.0. Because service implementations should only overtake type mapping tasks, also for performance reasons, it is discouraged from translating between different query languages, which should happen on a higher level in the application logic if necessary. The query operation returns the result of the query evaluation formatted as string. Note that this operation may eventually return an empty string, if the query could not be satisfied, but never **null**. In case the invoking party could not be properly authenticated, no document corresponding to the provided Id could be found or in case of a database error, a **StorageException** is thrown.

update

Analogous to the query operation, the update accepts an update statement according to the update mechanism provided by the underlying database. To stick with the examples provided previously, Apache XIndice and eXist [68] use the XUpdate language [162] and Berkeley DBXML incorporates a proprietary update mechanism. The update operation does not replace existing document contents, but modifies portions of XML document data. A **StorageException** is thrown in case of insufficient client authentication, an attempt to update a non-existent document is made or some internal fatal runtime error occurred.

6.3   Workflow Services

6.3.1   SemCrypt Service

The SemCrypt service provides access points to all available basic functionality adherent applications provide. Basic functionalities comprise all operations provided by the database management system, i.e. operations that can be performed on the database. In contrast to the storage service described in the previous section, the SemCrypt service also offers CRUD operations on schema information. Since implementation of schema operations is far from trivial and implies well sophisticated implementation details, this functionality has been omitted from the storage service directly operating on the database. The SemCrypt service should be considered as optional entry point for any SemCrypt application workflow that has been already implemented and tested and could thus be used as starting point for any XML based process definition, such as BPEL [91]. Concrete implementations should provide the functionality as specified within this service and access the core components through the other services as defined in the previous chapters. Operations on XML documents as well as the query and update operations are not thought to be mapped directly to the storage service, since these require authorization as well as encryption/decryption of storage data, which is only handled by the appropriate authorization and cipher services.

Optionally, an application may support CRUD operations for documents and corresponding schemes on role and user level as described by the SCDBF. These operations are thought to replace the general purpose operations and allow a wider variety of valid documents for an application implementation. However the implementation of those operations is an optional extension and does not adhere to the SemCrypt application workflows. Therefore invocation of those methods is not SemCrypt conformant and should not be implemented in conjunction with the DKE data store. The following table lists the operations of the SemCrypt service, including input parameters and return types. It should be noted that fault messages have been removed from this summary for reasons of simplicity. However, each of these methods may throw a **SemCryptException** message, which wraps any error that may occur during execution and reports it to the invoking party. Since this is the only port for reporting errors to the calling entities it has to be ensured that any exceptional states are handled appropriately and forwarded to invoking parties using the SemCrypt exception type. As already seen with the other service specifications, each method contains an optional token parameter, which serves as placeholder for any application level processing purposes, such as authentication. In the default implementation, this token is ignored and has no impact on the operation invocation.

insertDocument
Basic document insertion requires the document content, i.e. the complete content of an XML file, formatted as string and a string identifier, which must be unique throughout the application the document refers to. It is up to the invoking client application how this identifier is provided, whether it is automatically generated or assigned by the individual users comparable to document file names. Documents stored via the **insertDocument** operation are stored in the external insecure database. Metadata, such as referring policies, ciphering instructions, relevant schema or optional versioning in-

formation, are written to the document table on the secure data store as specified by the SCDBF. Since documents accepted by this operation do not specifically belong to a user or group of users, they do not explicitly refer to any subject in the database schema. The operation must throw a SemCryptException fault in case the identifier already exists, the user is not privileged to invoke the operation or some internal database error occurs during sensitive document metadata or external untrustworthy document storage.

selectDocument
Obtaining a document requires the invoking party to know the identifier of the requested document. The operation returns the document formatted as string only if the requesting user is the owner of the document or the user is entitled to see the whole document. A user is assumed to have access to the complete document, if an access policy exists, which grants the user access to the document root via the query operation discussed below. The operation must throw a SemCryptException fault in case the access to the document cannot be granted, the document could not be found or in case any internal exceptional state occurs during operation processing. In less restrictive application environments it is worth the consideration, whether all portions of the document are returned the invoking parties are allowed to see, based on the specified access control policies.

updateDocument
This operation is functionally similar to the insertDocument operation. Basically it should replace the old document with the new one. Documents, which do not exist in the database according to their identifiers, are not candidates for being updated. In this case the operation must throw a SemCryptException fault. Analogous, the insertDocument operation is not supposed to independently update a document, but indicate a fault if the provided identifier already exists in the security storage. Users allowed to perform this operation are document owners and role administrators respectively.

deleteDocument
Subjects allowed to remove documents are document owners, i.e. single users or role administrators. Access control is performed in conjunction with authentication on operation invocation. The operation must indicate a fault message if a non-existent identifier is passed on to the delete request, the invoking party is not allowed to perform the operation or some internal exceptions are raised during transaction processing.

insertDocumentForRole
This operation is a special implementation of the `insertDocument` operation and necessary to associate uploaded documents with a group of users. The subject attempting to perform this operation must be role administrator in order to successfully store a document for the group otherwise a `SemCryptException` fault is returned to the client. Since a user may belong to several groups, the role Id must be passed on to the operation to determine the role the document should be associated with.

insertDocumentForUser
This one is intended for single users to insert their private documents. What `insertDocumentForRoles` does for role administrators is solved by this operation, which allows even more specific data management.

insertSchema
Documents saved in the data store must adhere to a specific XML schema, which must be defined when using the DKE database management system, but is optional in other conventional XML databases. Thus implementation of the schema operations may be omitted when deploying the SCWSF in settings using databases different from the DKE database. The `insertSchema` operation is the very first operation that has to be invoked on the security server when deploying a new application environment. XML Schema holds the central information required for nearly any other artifacts. Policies, ciphering instructions, versioning history and finally documents refer to schema data. The SemCrypt environment only accepts a single schema, which must be carefully designed before application deployment, because future modifications have a tremendous impact on all the other artifacts and require substantial modifications on authorization, key management and document storage. Further invocations of `insertSchema` after application deployment must raise a fault message and lead to immediate operation abortion. A schema may only be inserted by the application administrator during setup time, not by any subject registered in the application during runtime. It accepts a schema name and schema content, both formatted as string parameters and throws a `SemCryptException` message if an unauthorized party tries to invoke the operation or a schema has already been defined.

updateSchema
Since it is highly unpredictable, which program state it may run into, it is strongly discouraged from ever call this operation during runtime. If the application schema is becoming obsolete for some reason, a complete re-

deployment should be considered. SemCrypt does not yet implement any notion of schema evolution, which makes it seriously cumbersome if not impossible to change the schema during runtime and apply the modifications to all affected artifacts and at the same time not run into any application inconsistencies. Furthermore, in some application settings, single users may have stored their private documents on the server and provided their own access control policies, which must not be changed in order to preserve user privacy. The operation signature is the same as with `insertSchema`, the difference is the purpose of replacing an existing schema.

deleteSchema
Like updating a schema, `deleteSchema` has been specified for the sake of completeness in order to provide full CRUD support also for schema information. And like updating a schema, `deleteSchema` should never be called during runtime.

selectSchema
Provided the correct string identifier, this operation delivers the XML content of the schema information deployed for the actual application environment. Though SemCrypt security does not depend on the concealment of schema information, schema operations are basically reserved to application administrators. It throws fault messages in case of unauthorized access, non-existent schema data or database access errors.

queryDocument
The query operation is probably the most important communication point with the document storage. It takes a query string containing a valid XPath expression and a document identifier referring to the document for which the query shall be executed. After successful authentication, the document identifier is investigated and determined whether the user has access to the desired document or not. Afterwards, XPath queries are parsed and forwarded to the authorization engine, which applies access control policies accepts or denies the query request depending on the policy evaluation outcome. If access can be granted the query is forwarded to the data storage interface, which finally executes the query and returns the result to the invoking client. The security check must therefore consist of three steps: authentication, checking access rights for the document and query evaluation. Since all this implementation efforts aim at providing a secure database management service, querying and updating the data store is the main functionality SemCrypt application environments provide to client applications and will thus be most frequently called during runtime. Exceptions may occur due to authentication errors,

access control on documents and document fragments and technical issues during communication and request processing, and are reported by returning a `SemCryptException` fault.

queryDocuments
This operation works exactly as the `queryDocument` operation, but accepts an array of document identifiers to query more than one document in the data store and thus speed up query processing by eliminating unnecessary communication overhead. Otherwise the contract specified for `queryDocument` also holds for this operation.

updateDocument
Currently, the update of a document is triggered by prividing an XUpdate statement. Since XUpdate is a proposal, which development seems to be frozen and is likely to be replaced by appropriate XQuery profiles in the future, the current implementation of the `updateDocument` operation is likely to undergo a comprehensive revision in forthcoming releases. Nevertheless, the signature is similar to the `queryDocument` operations, with the difference that an XUpdate statement needs to be passed on instead of an XPath expression. Since each XUpdate expression contains an XPath statement to identify the portion of XML to modify, the operation needs to extract the statement and pass it on to the authorization module. The update procedure does not return anything, but eventual exception messages as already discussed with the query operations.

6.3.2 The Management Service

Apart from database operations, the SCAF provides interfaces responsible for handling application relevant functionalities, which are not covered by the other services. This applies to user management, key management as well as policy, ciphering and versioning definitions. User management is part of the SCDBF [143], key management comprises encryption/decryption processing as well as issues including dynamic key change and is discussed in SCEDF [144] and access control policies are subject to the core authorization framework and as such, specified within the SCAAF [142]. This section provides a list of operations offered by the application management service. Policy and ciphering information may be inserted, updated, deleted and selected and defined on a general purpose level, affecting all users registered in the application, on role level, overriding the general purpose instructions and being only valid for a specific group of users and on user level as well, overriding even role level and allowing very specific policy and ciphering definitions. As defined in the SCDBF, each document stored by the application, has a reference to its schema and may optionally hold a specific access control policy, encryption/decryption instruction and versioning document. The difference between the general purpose CRUD operations and those defined for users and roles is that users need to pass on the document identifier for gaining access to the specific file definition. Note that again, authentication information is omitted in method signatures. Depending on whether access to role or user specific documents is requested, the appropriate operation needs to be called, which in turn performs a different authentication process. On role level, the authentication engine needs to return the set of roles the user belong to, on user level it is sufficient to check whether the user is entitled to perform the specific operation.

insertPolicy

Policies are used in SemCrypt applications to restrict access to portions of XML documents identified via XPath statements. In each application at least one policy must be defined, which is valid for any document stored in the untrustworthy database. This policy must be added using the `insertPolicy` operation, which accepts a string containing the complete XML policy document and may only be called once during application setup time by the application administrator. Access permissions specified by the general purpose policy may be overridden by document specific policies, which may be added by single users or roles by calling the `insertDocumentPolicy` operation, discussed shortly. No users or roles may access this operation, which is indicated by a `ManagementException` message if any unauthorized party tries to perform this operation. Faults may also be triggered if a policy already exists and an administrator tries to access this operation or any internal error occurs.

deletePolicy

Removing the general policy is only a good idea in low-security applications or if access permissions are expressed via document specific policies. Since no Id is required or explicitly assigned to the single, general purpose policy, this operation does not require any parameter to identify the correct policy file. A a `ManagementException` is raised if a subject different from the system administrator tries to access the operation, no policy file has been specified so far, or any internal failures occurred.

updatePolicy

Implementing the update process of a policy is a comparably simple task, since no further artifacts involved in the application are affected. It simply replaces existing access privileges by some new one. Also, calling this operation is only administrator business. Fault messages are indicated by the a `ManagementException` type and are returned if the calling party is not member of the administrator role, if there is no policy to update, or in case of technical errors.

selectPolicy

To complete the CRUD operations for general purpose policies, it should be possible for application administrators to check the content of current deployed access permissions and restrictions. As with `deletePolicy`, selecting the one and only general purpose policy does not require any parameters being passed to the service operation. Unauthorized invocation attempts, a non-existent policy or internal service errors are considered as exceptional application states and indicated by a `ManagementException` message.

insertRole

A role is at least defined by a role name, a description and an optional set of base roles used for permission inheritance. Additional information required depends on the application scenario. Role information is wrapped by a role object, which must adhere to the role table as specified by the database schema on the security server. Adding a new role does not work, if a role with the specified role name already exists or accessing the database due to technical reasons or insufficient access privileges. In these cases a a `ManagementException` message specifies the reason of failure.

updateRole

Whenever a role needs to be updated, the `updateRole` operation needs to be called. Similar to the a `insertRole` operation, it accepts a role object with the required attributes. The role name must be the value of an existing role, other values are used to overwrite existing ones, i.e. description or base roles. A a `ManagementException` error is raised if no such roe with the specified name exists, or accessing the database failed for some technical reason or because of missing authentication credentials.

deleteRole

Removing a role may be accomplished by simply specifying the name of the role formatted as string. When deleting a role, relations to users and other roles as well as artifacts, specifically associated with the role, i.e. documents, policies, versioning histories and encryption strategies may be removed in a way no inconsistencies can occur as well. Removal may fail with a a `ManagementException`message if removing the role or directly associated

artifacts fails for internal, technical reasons, authentication failed, or no role
with the specified name exists in the database.

selectRole

By providing the string formatted role name, the complete structure of a role
may be returned to the invoking client. This involves at least description and
the identifiers of eventually specified base roles, but no artifacts associated
with the role, such as encryption keys, documents or policies. This operation must throw a `ManagementException` if any unauthorized party tries to
access role information or no role with the specified name exists. Database
failures must be reported as well.

insertUser

Each application may define a different user data structure, depending on
which user information need to be stored. The operation only accepts a user
object, which must adhere to the user definition of the SCDBF user table.
It must contain a unique identifier, such as a username, a customer Id or an
E-Mail address to uniquely obtain user information from the database and an
optional set of roles the user belongs to. It may throw a `ManagementException`
message if the user Id provided already exists, the requesting party is not
entitled to perform the operation, or accessing the database failed for some
technical reason.

updateUser

Modifying user data is only possible for user information already existent
in the database. Updating should be basically possible for any user data,
except for unique usernames. The operation accepts a user object as parameter, which must contain the user Id to gain the correct reference to the
user object in the database. All the other attributes may either contain no
value, which indicates that no modifications need to take place or any value
different from the existing as long as the data type is valid according to the
database schema. The operation may raise a `ManagementException` fault
if no user with the Id provided exists, the requesting party has insufficient
access rights, or update failed for some internal, technical reasons.

deleteUser

In order to permanently remove a user from the application, only the unique
identifier is required. Deleting a user involves not only removal of user information, but also any information associated with the user, including documents, policy files, ciphering instructions and role associations. If the last
user of a role is removed, removing the role may also be considered. If a

user is removed, who is at the same time role administrator, another representative should be specified. However, these issues may be implementation dependent and is currently not supported. The operation must throw a `ManagementException` message if no user with the Id provided exists, the requesting party has insufficient access privileges, or in case of technical failure.

selectUser
Retrieving user information happens by passing the user Id to the operation and obtaining a data structure containing the user information, just like the one passed to `insertUser` and `updateUser` when writing user information to the trusted data store. Invoking the `selectUser` operation is straight forward and only fails with a `ManagementException` message if the requesting party is not entitled to perform the operation, no user with the Id provided exists in the database or some technical error occurred during processing.

insertKey
A cryptographic key in SemCrypt environments is made up of a single unique identifier and the key itself formatted as array of bytes. When inserting a new key, the identifier passed on with the invocation needs to be checked for uniqueness and writes the key information to the trusted key store. The identifier must be unique throughout the application, since users and roles do not have a separate key store to save their keys. The key identifier is the same value that has to be used by the XCipher encryption/decryption instructions as stated in [144]. Providing cryptographic keys at the secure server is entirely optional in each SemCrypt based application, since users may decide to store their cipher keys locally on their private machine. In this case the user either knows which key to apply to the encrypted data, or obtains the ciphering instruction from the server by invoking the `selectCipher` operation from the `ManagementService`. In case a user performs a decryption request on the security server, the cipher module attempts to obtain the correct cryptographic key as specified in the ciphering instruction from the key store to decrypt the data or throws a fault message if it fails to do so for some reason. The `insertKey` operation raises a `ManagementException` fault if a key with the Id specified already exists, the requesting party is not allowed to perform the operation or if some technical problems are encountered during processing.

updateKey
Updating a key is tries to replace an existing key with a new one. However, this operation should be implemented and used with great caution, since it has a significant impact on document data involved in the application.

Dynamic key change means that affected encrypted documents need to be obtained from the unstrustworthy, external data store, decrypted and then re-encrypted with the new key. Furthermore, if the affected key is not only stored at the security server, but also at client hosts, appropriate propagation of the key change has to be taken into account when implementing this operation. In current SemCrypt settings a key is assumed to remain constant throughout an application lifetime, i.e. it cannot be changed, which in turn makes the current default implementation to ignore this operation and throw a `ManagementException` fault on invocation to indicate the missing implementation. When this operation is implemented, a `ManagementException` message must be returned to the invoking parties to indicate missing authorization, a non-existent key or technical problems on the server side.

deleteKey
Removing a key from the key store is somewhat cumbersome as updating a key, since it may affect documents and ciphering instructions used in an application setting. Deleting a key involves decrypting affected documents and removing corresponding information from the ciphering instructions leaving document data being written back in plain text to the untrustworthy data store. This has a serious impact on the security level an application may assure to interacting users. The operation takes the unique key identifier as parameter to permanently remove the key from the key store and triggers ciphering and database access components to obtain and decrypt sensitive document data affected by this operation. A `ManagementException` fault must be raised whenever authorizing the invoking party fails, no key can be identified by the Id provided or any technical failures occur during processing.

selectKey
This operation returns the array of bytes containing the actual key information from the security key store. The unique string identifier of the key must be provided as parameter. Selecting a key has no further impact on any internal application states on the security server. Like the other ManagementService operations, `selectKey` throws a `ManagementException` message if authorizing the invoking party fails, no appropriate key could be identified by the Id provided as parameter, or technical problems occurred.

6.3.3   SemCrypt Security Services

The component discussed by the SCAF [140], are general purpose modules with a well-defined API to facilitate application integration and to enable enhancements independent of the actual configuration. However, SemCrypt

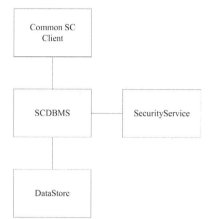

Figure 6.1: Basic SemCrypt component interactions

originally aimed at very specific application settings with very specific work-flow definitions. This envisioned the deployment of an arbitrary number of untrustworthy data stores holding tons of binary data, which semantic meanings could only be recovered by the deployment of a suitable and specifically tailored security service. This service and possible access mechanisms shall be discussed within this section. In typical SemCrypt application scenarios, we differ between untrustworthy storage providers, trusted security servers, which perform query and update authorization as well as storage unit encryption and decryption, and trusted or at least semi-trusted clients, representing the user application front-end. These architectural differences are discussed in detail in [141]. However, the requirements for the security service are independent of the concrete architectural setting. The difference is located in the interaction between the SemCrypt Database Management System (SCDBMS) and the security server. The client application provides query and update as well as document upload and modification functionalities and passes required data on to the common SemCrypt client component, which provides a uniform view of the underlying database management system to the invoking client application and thus facilitates end-user application development. The SCDBMS is responsible for performing any pre-processing operations on the data provided, including document indexing, query processing and metadata management. However, each path that is affected during query optimization needs to be authorized by the security server, before being executed on the data store. The resulting data is sent back to the security server for decryption and should be eventually filtered

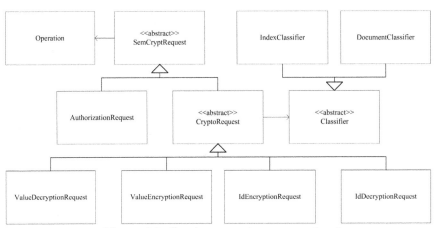

Figure 6.2: Service request message structure

to ensure that no unauthorized data is being sent back to the client application. In case of an update operation encryption is performed immediately after authorization has been successful.

For accessing the security server, the SemCrypt Security Services Framework (SCSSF) [147] provides a Web service interface for easy integration with the SCWSF and TCP and RMI interfaces for performance issues. Queries may be executed very quickly and in short time intervals, which makes a Web service susceptible to performance bottlenecks due to expensive SOAP parsing. The Web service interface is different from the RMI and TCP interfaces with respect to how authentication information is passed on from the client application. Web services as specified by the SCWSF and corresponding component documentations take advantage of incorporating WS-Security headers within SOAP message to communicate required security issues. Therefore, authentication information, such as username and password are transmitted implicitly within each SOAP message and handled by the application server hosting the service. Neither RMI nor TCP technology provide some built-in security mechanisms allowing to implicitly pass required security information. Authentication has to be performed on the application level, i.e. the implementation of the RMI/TCP services respectively. For that purpose both RMI and TCP services are defined on the basis of a common message structure, which is illustrated as simplified UML class diagram below: Figure 6.2 illustrates the request message structure used by the SemCrypt security service. Supported messages are authorization, value encryption/decryption and id encryption/decryption requests. All messages must contain a username, a password and a desired operation to be performed. Currently operations comprise basic CRUD functionality. An authorization request additionally contains a path value, which must be a valid XPath 2.0 statement, which is used to perform basic authentication checks. Detailed analysis of the path expression is not required, since query processing and optimization is performed at the SCDBMS and sub-queries must be handled by the security server subsequently through id or value encryption/decryption requests. The basic authorization request is optional and if used without further requests must guarantee correct results, i.e. grant or deny access. Different authorization approaches are discussed in SCAAF [142]. Any encryption/decryption request must contain a classifier containing the path value information, represented by either an index or a document classifier. The difference in both classifiers is how the path value, i.e. the XPath expression is provided. The document classifier contains the path value and the id of the affected document the same way as the authorization request does. An index classifier denotes an index value, which is used to obtain a path value from the metadata manager. Hence, there are two equivalent possibilities to retrieve a path value and authorize it, either by directly passing it on to the security server or letting the server obtain it from the metadata manager.

Classifiers indicate how path information is passed to the security service. Path information in turn is used to perform authorization and obtain the correct key for encryption/decryption of sensitive data from the secure key store. Applications may support encryption using multiple cryptographic key, which makes it necessary to remember which parts of XML data has been encrypted with which key. This is described in greater detail by the SCEDF [144]. Ciphering may be performed directly on data fragments or id values. Ids are again subject to metadata management and do not affect the security server, which must only take care of authorization and ciphering. What kind of data is being encrypted or decrypted is nothing for the server to worry about. Nevertheless, this needs to be differentiated with respect to the message structure. Figure 6.3 illustrates the response messages structure analogous to Figure 6.2 representing the request messages. The diagram should be self-explanatory. Value and id encryption/decryption responses only contain an array of binary data containing the encrypted/decrypted data or **null** if something went wrong, mainly caused by failed authorization or ciphering process. The authorization response carries an information message indicating whether the authorization process was successful or not. The following list gives an overview of the data types used for representing the message structure. It specifies parameter types and a brief description

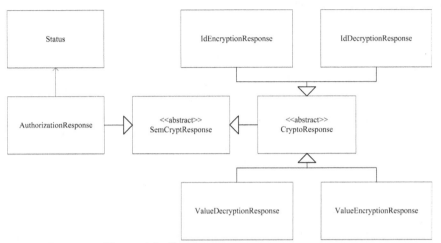

Figure 6.3: Service response message structure

of their purposes.

SemCrypt request and response classes represent base request/response structures. A request must always contain a username and a password as well as an operation value containing the type of request. An authorization request additionally contains the path value denoting the XML resource attempted to being accessed and a document Id identifying the document the access should be applied on. The authorization response holds the authorization decision, i.e. whether the access has been granted or denied, and a textual description of success or failure respectively. An encryption request may be either made up of a value holding the plain text to encrypt or an Id, which is used to obtain the value from the metadata manager [93]. Responses simply contain the encrypted value as byte array. Decryption may be requested by providing directly the cipher-text or an Id identifying the cipher-text in the metadata manager.

These message types are used for the RMI and TCP interface. The TCP service runs a thread opening a socket for each incoming request, then determining the concrete message type and forwarding it to the message handler. Errors are indicated by returning appropriate information within the response messages as presented above. With RMI, methods on the server are called, where each method accepts exactly one parameter, i.e. a message. Throwing method specific exception type in case of an error is not specified, since clients interacting with TCP sockets should expect the same server behavior when changing the service interface to RMI. Return values should be handled equally in both cases, such that everything that needs to be changed is the client interceptor performing the invocation. The operation signatures should be straightforward and are listed below: Both the RMI and TCP

	Parameter	Return value
authorize	`AuthorizationRequest`	`AuthorizationResponse`
encryptValue	`ValueEncryptionRequest`	`ValueEncryptionResponse`
decryptValue	`ValueDecryptionRequest`	`ValueDecryptionResponse`
encryptId	`IdEncryptionRequest`	`IdEncryptionResponse`
decryptId	`IdDecryptionRequest`	`IdDecryptionResponse`

Table 6.2: Security operation signatures

interface use username/password information with each method call to perform authentication checks, since no session handling mechanism is provided. Authentication methods are discussed by the SCAAF in [142].

With Web services, the interface definition needs some little modifications for authentication reasons. As discussed in [148] the SCWSF communicates security information through the SOAP header, which at least also includes authentication information. With Web services, authentication information, i.e. username and password are rather implicitly transferred to the application server and handled by the underlying SOAP engine than explicitly as method invocation parameter. Thus username and password are being extracted from the request messages presented previously and the remaining parameter values are transferred separately to avoid unnecessary data transfer, since it will not be handled anyway. Adapted service operations are listed below. The full service interface can be found in Appendix B.

authorize
Signature and behavior of the authorization operation are equivalent to the authorization operation of the authorization service. What is needed is a path value identifying the protected XML resource, the desired operation and the document identifier, the request is targeted at. All these parameters are formatted as strings. No return value is expected, instead a `SecurityException` error is raised if authorization fails. This operation is intended as replacement of the authorization service operation, but was added to the security service to logically group all operations relevant for the service functionality. Implementation may either simply forward the request to the authorization

service, if one is deployed, or directly access the authorization module on its own behalf.

encryptValue

For value encryption a classifier must be provided to indicate whether the value parameter already contains the value to encrypt or only provides an identifier, which must be used to access the metadata storage in order to obtain the value from there. When calling this operation, authorization is performed implicitly requiring the desired operation being passed along with the classifier. If the procedure completed successfully, the encrypted value formatted as byte array is returned to the caller. A SecurityException error must be raised in cases authorization or encryption fails or insufficient authentication information has been provided. The error message thus wraps potential AuthorizationException or CipherException messages, if such an error occurred.

decryptValue

This operation represents the counterpart of the encryptValue procedure. Parameters and error message are just the same. Instead calling the encryption operation on the cipher service or cipher component, the decryption operation must be called analogously. The return value of the decryption operation is therefore the decrypted vale formatted as byte array.

The procedures encryptId and decryptId work just like those operating directly on the document values. Ids identify the corresponding values in the DKE Linz metadata store.

Nevertheless, the client code may remain widely untouched, since no specific exceptional messages are thrown and handled by the client. The request interceptor at the client side only needs to extract authentication information from the message, add it to the SOAP header and pass the rest on separately. Since most data types in use are basic types, mapping from programming language specific type to XML type is trivial.

Chapter 7

Proof of Concepts

The previous chapters were focused on implementation details of the server-side core components. The following sections deal with the presentation of a development environment for applications based on SemCrypt technology and security service provisioning on the client side. In concrete, the first section documents an Eclipse Plug-In, allowing basic SemCrypt application design and deployment in terms of access control and data encryption, and an automatic Web user interface generator based on XForms, which may optionally be integrated with the Plug-In extension. The user interface generator was developed in the course of a master's thesis with the purpose to investigate current state-of-the-art Web technologies and their applicability regarding the generation of device-independent user interfaces. Functionalities comprise querying and modifying database content via a form-based Web application, which in turn translates user selections into valid XPath and XUpdate statements and transmits them to responsible services for further processing. Furthermore, several client applications, which take advantage of the SemCrypt services, shall be presented in detail: The first application was created for both testing service functionalities and demonstration purposes of SemCrypt capabilities. The application was made available to students at the Vienna University of Technologies, who were supposed to model their competencies in a university lab using a subset of HR-XML and store their documents on a remote server, which could then be queried and modified. Other applications presented are a secure SMS data store with focus on security management on mobile devices and collaborative project management using Microsoft Excel.

7.1 Development Environment

Chapter 6 provides a detailed specification of the service interfaces, which can be referred to as programming interface in cases programmatic integration with existing application systems is required. For developing new applications from scratch, the SemCrypt toolbox comes with a graphical user iterface, which may for instance be used as Eclipse Plug-In [19], to develop simple SemCrypt based applications via drag and drop. This and an extension allowing the generation of Web based user interfaces, which enable direct interaction with the SemCrypt services, are subject of the following sections.

7.1.1 Eclipse IDE Integration

In order to facilitate application development on the basis of SemCrypt technology, a Plug-In for the Eclipse IDE platform has been developed. The Plug-In is designed as graphical user interface, which is composed of a set of dialogues, each with a different purpose. It currently supports the deployment of basic SemCrypt functionalities, which are listed below:

- Schema import and document validation

- Creation of users and roles

- Definition of a general purpose XAccess policy

- Definition of a general purpose XCipher encryption strategy

- Optional deployment of Web interfaces (see next section)

The first step that needs to be taken when developing a SemCrypt compliant application is to define a schema, which describes valid XML document structures for an application. The GUI allows importing only one schema file, but is capable of resolving references to external schema files automatically. Since only one schema is allowed per application, internally, a single schema is generated, which is used in the application thenceforward. After schema import, it is possible to add an arbitrary amount of XML documents, which are automatically validated against the newly created schema file. Afterwards, the application developer is asked to define a set of users and roles that are going to be registered in the application setting.

Based on the schema file generated at the beginning, the Plug-In interface generates a complete list of XPath statements, which are allowed for documents associated with the application being developed. These XPath expressions (compare Chapter 2 and Chapter 5) have the form:

```
XP := ("/"(N":")?E)* ("/"(N":")?  "@"A)?
```

The XPath statements are displayed as tree, containing elements and attributes specified by the schema file. When selecting a branch or leaf in the tree, additional information about the selected node, which is extracted from the schema file, is displayed in the Plug-In window. This information comprises node comments, data type of the selected element or attribute and a Boolean flag whether the selected node may contain text or other element nodes. Data types provide indications of what kind of data a node may contain and assist the application developer if they wish to restrict access to a node with respect to the value it contains. For instance, referring to the schema file in Appendix A, we can easily determine that the element CompetencyId is made up of three attributes, namely id, idOwner and description, each of built-in schema data-type string. With that in mind, we may easily restrict access to any element CompetencyId, which attribute id contains the value ICT, without having to open the schema file and checking data-type and allowed barriers ourselves. Of course, this only works with simple types. Complex types may define sequences of other elements and attributes, which in turn may specify their own data-type, which makes it hard to define restrictions for complex schema types. Indications about whether an element may contain other elements, attributes or text is especially interesting in cases we wish to restrict access to a certain sub-element, but not to the textual content, for instance. When selecting a node in the tree representation, node information is automatically updated and an input dialogue opens, which allows defining access control rules for the node selected. Access control rules embrace the definition of conditional access control rules, i.e. value barriers as just explained and access rights. Access rights are actions that may be performed on the selected node. Since SemCrypt provides an application framework for secure storage provisioning, actions supported by the Plug-In are basic CRUD operations, i.e. insert, select, update and delete. To clarify the access control definition process, Figure 7.1 illustrates the corresponding interface provided by the SemCrypt Eclipse Plug-In.

This allows easy specification of basic permissions for an application, without asking special knowledge of the XPath language from the application developer. After defining the access control rules, the Plug-In internally generates XPath statements, compatible to the grammar defined in Chapter 6, and creates the general-purpose policy file named xaccess.xml, which is used

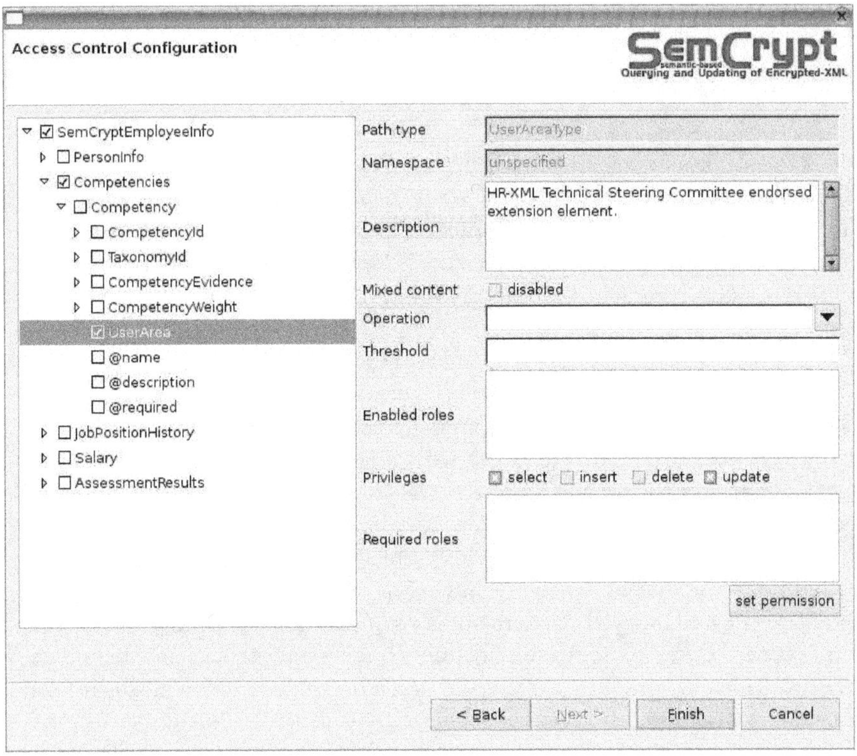

Figure 7.1: A screenshot of the Eclipse Plug-In taken on Ubuntu Linux

by the SCAAF after application deployment, which can also be managed by the Plug-In. Nevertheless, the Plug-In is intended as supporting tool for application developers, who are just making their first steps using SemCrypt technology. In its current state of development, the Plug-In is nothing more than a prototype, which was also made for testing and evaluation purposes, and surely has some limitations, which shall be discussed shortly.

After finishing permission definition, an XCipher encryption instruction may be specified as well. The Plug-In currently does not support the generation of cryptographic keys, but rather prompts the developer for a local key-store to obtain existing keys from. If no such key-store exists, the reference implementation generates a set of dummy keys by relying on the local Java Cryptography Extension (JCE) [157]. If this does not suffice, the developers have to take care of key generation and provide a key-store themselves.

The XCipher window displays the same node tree as known from the access control configuration, but adds different, encryption related, options. As explained in Chapter 5, ciphering configuration does not provide such fine granular XPath syntax as XAccess does, but the idea is the same: On selection of a node in the node tree, the developer may specify an arbitrary number of cryptographic keys by their identifiers, which are also used for the local key-store, that should be used for the encryption sequence of an XML sub-tree.

The final configuration step is optional. It integrates the possibility to let automatically generate a user-specific Web interface. Each user or each role, respectively, has different permissions in the application environment. The Plug-In uses the XAccess permission definition to generate an XForms based interface, which provides exactly the functionality, the user is allowed to access. More information on that is provided later on in this chapter. Application designers are asked to provide some technical parameters, which are necessary to deploy the Web application, including Web server URL and database server attributes. Future efforts are focused on stability testing and enhancement of existing functionality. So far, basic application deployment is supported. Due to the complexity and openness of the SemCrypt application framework, a list of possible extensions are considerable, such as including update restrictions, user-defined access rights, user and role specific policies and encryption instructions, more sophisticated document management as well as automatic application deployment and runtime monitoring and configuration options.

7.1.2 Web Application Generation

XForms [32] is a promising and emerging standard for describing Web forms in a presentation neutral manner, independent of platform or input device. XForms is XML and thus describes meta-forms which can be used to generate concrete user interfaces, such as XML related languages, such as HTML or WML, procedural formats, like PDF or PostScript, and many more. Since XForms is abstracted from concrete presentation layout it requires a processor for rendering the output on the underlying device. This processor can be either deployed on the server [43] or on the client, for example as Web browser Plug-In. XForms has been chosen, because it is the result of a W3C standardization effort and has the advantages of being open source and platform independent. Competitive products are Microsoft XAML [111], XIML [183], UIML [165] and Mozilla XUL [116]. UIML is standardized and platform independent, but only partially free and open source and requires an own browser

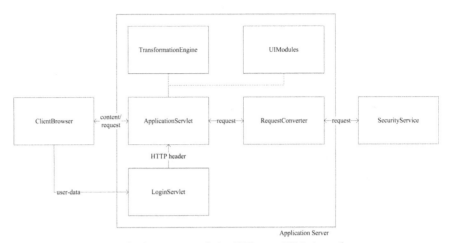

Figure 7.2: Architecture of the XForms Web interface generator

for being displayed. XUL is free and open source, but not standardized and only supported by Mozilla browsers. XAML runs only on Windows Vista and Internet Explorer and is a commercial product. The term client may be a bit misleading in the context of the application in hand, since it requires a separate application server to run, service consumer may be more appropriate, nevertheless it is considered to be a Web service client. When a user logs in via the login Servlet, authentication is performed using the SCAAS modules and if successful the request is forwarded to the application's main Servlet. Required content type, such as XHTML or WML as well as preferred language are extracted from the user's HTTP request and used to determine the transformation module most suitable for the client device. Since XForms pages must be embedded within existing documents, this step is required to decide which XSL transformation must be used, including style-sheet information. Currently, this client provides HTML modules for Novell Internet Explorer and Mozilla FormsPlayer [71] Plug-Ins. Actual XForms rendering components are not illustrated, since they may be either be deployed on the client side as Web browser add-on or as server-side module. In the latter case the Web browser only performs client-side data validation. The following image (7.3) illustrates the result of the interface generation process: The layout of the user interface is defined by Apache Velocity [160] templates actually generating the XForms markup. A user interface contains both generic and fixed content. Fixed content are namespaces for XForms elements and Events, basic XML schema data-types and a header including title, icon and copyright. Other elements, such as navigation, form submission and input

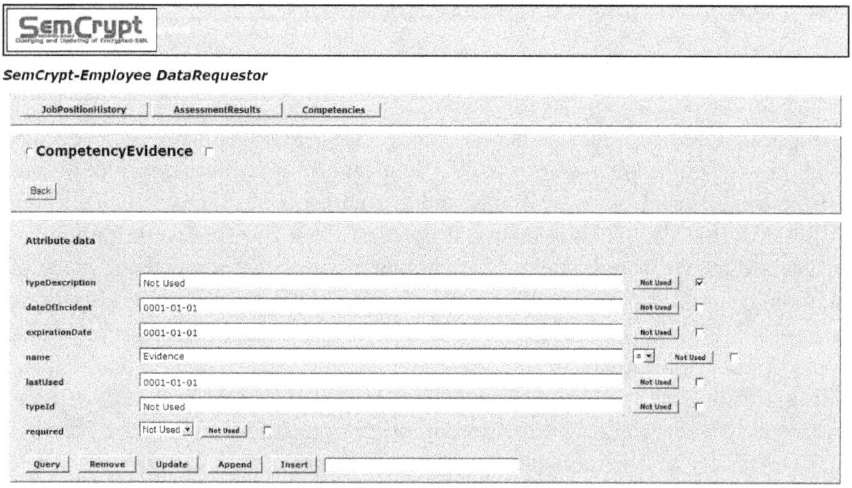

Figure 7.3: The generated default XForms user interface

handling remain generic. The interface is structured hierarchically, starting from the root element of the document data to be queried. Via radio buttons, child elements of the current context node may be selected, which triggers an event to reload the page with the selected element as root. This allows navigation deeper down the tree hierarchy. Element and attribute values may be further restricted via appropriate form elements, such as text-fields, checkboxes or drop-down lists. If a selection has been made, the user may state a selection or update request, which makes the Servlet pass on the selected content to the conversion module, which in turn generates XPath or XUpdate statements respectively. From then on the SemCrypt security services are invoked as usual: Authorization is performed and if successful the desired operations are preformed on the data storage.

It should be noted that information for user interface generation is not taken from the application specific XML schema, but rather from the XAccess policy files. When a user logs in, corresponding permissions are loaded through the XAccess parser and returned to the interface generator. This prevents generation of content, the user is not allowed to access anyway. Restrictions are handled as well. Nevertheless, the authorization service should be contacted further to rule out the possibility that generated XPath or XUpdate statements have eventually been intercepted and modified.

7.2   Use Case Implementation

7.2.1   Human Resource Management

Developing a security service is one thing. The other is to provide a service, which does what it is supposed to do. This should be proved by making basic functionalities publicly available and letting users interact with a concrete application scenario. This scenario was taken from the academic human resources sector. In a university lab, students from the Vienna University of Technologies were asked to model their personal data and competencies using a subset of the HR-XML standard and store it on some remote server using a Web application interface (`http://move.ec3.at:8180/cms/`). Apart from testing purposes the application was developed with the intention of demonstrating system capabilities to potential industry partners as well as to the Vienna University of Technologies career center (`http://www.tucareer.com/`) for further deployment. If users trust the security of the system, which was the assumption, then they would start using Web services in the literal meaning more intensively even with sensitive private data involved. Users could store their competency data externally, which could then be queried by potential employers, who are looking for young academic personnel.

Regarding the application deployment some modifications with respect to the default deployment process of the SemCrypt Application Framework (SCAF) were made: Authentication as included by the SemCrypt Authentication and Authorization Framework (SCAAF) was replaced by the authentication service provided by the Vienna University of Technologies to ensure that only students have access to the system using their university passwords and to retain scalability and reusability when demonstrating the system in future university courses. Note that students need not be employees at the same time, hence the `SemCryptEmployeeInfo` schema from Appendix A might be a bit misleading. Nevertheless, the main focus lies on competence modeling and employment histories and salaries may be optionally provided as well. Furthermore, a separate login procedure for academic staff of the university and business partners was created. Exemplary, these authentication processes are shortly demonstrated by the student login procedure.

Despite this set of modifications, services functionalities and security capabilities could be successfully tested, since the most crucial components, including authorization and the encryption/decryption module were deployed as well. The DBMS provided by the DKE Linz still serves as data storage holding encrypted sensitive data as discussed in the Security Services section

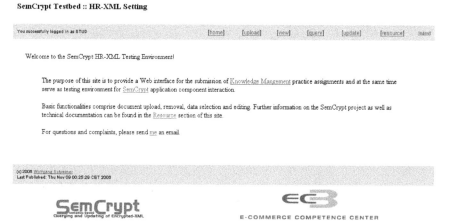

Figure 7.4: The HR-XML Web application interface

in Chapter 6. Unfortunately, many attractive features, including performance tuning did up to now not have the resounding effect due to moderate user participation of about thirty students over several weeks. Figure 7.5 illustrates the involved architectural entities and their initial communication.

When a student wants to login, he calls the Web application via a Web browser, enters student identification number and password, which is then forwarded by the application server to the Vienna University of Technologies authentication service along with a unique application identifier. This Id tells the authentication service to which URL the user should be redirected if authentication was successful. Note that communication must be carried out over SSL, such that the application cannot see the user password. If authentication succeeds, the authentication service generates a unique session token using student Id, password, client hostname and current timestamp and sends it in addition with the Id back to the application server. The reason why there is an additional Web server is involved is that whenever the URL of the application changes, it is not necessary to tell the university, but only modify it on the private server. The application server performs token validation and if successful, an HTTP session is started and the user may invoke the application services.

For academic staff and external partners a different user interface is provided, which limits user permissions to read access. Since the documents belong to students it did not seem reasonable to allow external users to mod-

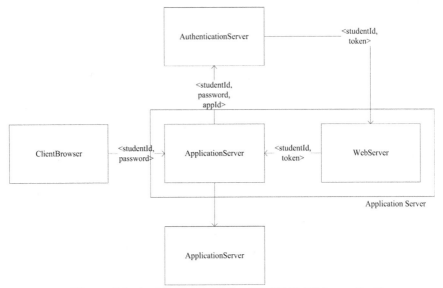

Figure 7.5: Architecture of the HR-XML Web applications

ify the content. Queries however are checked by the authorization engine before forwarding them to the data store. Decryption of the database content is automatically handled, since the keys are known by the trusted service domain in this setting. Similar to students, teaching staff can be authenticated by the university authentication service as well, for external users, such as companies, accounts are created on request.

In concrete services comprise document upload, query a document using XPath or modify it using XUpdate, as shown in chapters 2 and 3. The screenshot 7.4 shows the main page of the Web interface. In the HRM application first insights into the usage of the technology could be received. Privacy concerns and the difficulty to evaluate their own competencies were evident problems here. However, the technological problems were solved in the first prototype. In this application the framework could be extended in such a way that students can define their own access control assertions, since each student will have different security concerns. This, however, means a potentially very large number of assertions which could lead to performance problems. Techniques from recommender systems may help here.

7.2.2 Distributed Project Management

Artifacts, which are created during the lifecycle of a (software) project, often contain confidential information, which may only be seen by authorized people involved in the project. Sensitive data may comprise program source code, innovative research results or budget related information. Mainstream applications, such as Microsoft Excel or Project are widely used for electronic support during project settlement. A prototype developed in the course of a master's thesis [64] shall evaluate the applicability of SemCrypt technologies in combination with existing client-side collaborative applications. Project management with Excel is often performed in SMEs, which often do not implement satisfying security policies. Excel files, which have a corresponding XML schema could thus in the future be stored encrypted using the Sem-Crypt security infrastructure.

7.2.3 A Secure SMS Storage Provider

Lorry drivers often receive delivery orders on their mobile phones via SMS. In some cases it may be convenient to store order information remotely at a secure storage provider, which may be queried later on ordered by time, location or subject. After delivery the storage may be updated by deleting accomplished orders. With that scenario in mind the client application should implement a fat client with ciphering components directly implemented on the mobile device, in case messages are already sent encrypted to the drivers [66]. This application should only incorporate remote authorization, since access control needs to be performed in a trusted environment, but keep cipher keys in a local key store. Challenges to investigate encompass typical queries and updates, device hardware persistency and performance analysis with respect to remote service access, local key management and supported encryption/decryption algorithms. Prototyping includes JavaME [158] on Java compatible mobile phones, Blackberries, Palms or Pocket PCs.

7.2.4 Document Archives in E-Tourism

Customer care is one major key factor of success in E-Business. In this sector, service orientation gained tremendous significance to stay competitive. Companies have to be innovative and flexible to satisfy customer needs at the best possible rate to win their loyalty. In Part II, E-Tourism has been identified as promising fields for establishing new E-Services. In a bachelor thesis, carried out at the Vienna University of Technologies, an XML schema has been developed, which defines information structures describing tourist destinations as well as tourist profiles allowing local tourism service providers as well as travel agencies and accommodations, statistical analysis of tourist data and thus adapt their services in a customer-oriented manner. Particularly latter type of information needs to be protected, since it may contain personal sensitive data. The schema relies on a subset of the Open Travel Alliance (OTA) XML [161] specification. Target of the thesis was to analyze potential application scenarios, typical query and update structures, groups of users involved and suitability of tourism scenarios for a variety of end-user devices.

7.2.5 Encrypted E-Mail Storage

Another use case of sensitive document archives may be identified by an untrustworthy E-Mail storage provider. Sometimes people may wish to securely archive their private E-Mails on some external storage provider or wish to encrypt their IMAP account, since they do not fully trust their Mail provider. This scenario could be very well extended to Web hosting services, allowing to secure parts of a Web site and making it accessible only to selected users, for instance. The application has been designed as single-user scenario, which makes use of the security components discussed within the thesis, but does not require resource locking and multi-user control. E-Mail messages are translated into an XML dialect to allow additional storage of metadata information and to take advantage of the SemCrypt framework features.

7.2.6 A Generic SemCrypt Testbed

The final prototype discussed herein is not an application setting per se, but rather a generic environment for demonstration purposes of SemCrypt storage features. The tool is available both as stand-alone and Web application and provides a GUI frontend, which visualizes internal data storage processes. For each user, who registers in the testing environment, a new data store is created, which supports arbitrary schema and document upload as well as modification operations and XPath and XUpdate processing. Authorization and encryption processing are integrated as well. The focus however lays on the graphical representation of technical details of the database internals. The tool has been made available on the SemCrypt project Web site [177] for public usage.

Part IV

Epilogue

Chapter 8

Conclusion

An article by Thuraisingham [163] briefly discusses security and privacy concerns in the Semantic Web, E-Business and knowledge management and their impact on data mining and information extraction. Security mechanisms, for example authentication or encryption, are considered essential and must be incorporated into all aspects of (semantic) business processes.

Trust management and flexibility of access and usage control policies are indispensable for future research efforts on developing a security framework for semantic E-Business applications. Lee et al. [99] consider knowledge management as one key factor for economic success. They state that if a company's knowledge is not properly protected, the company will lose its competitive advantage. Since the Semantic Web widely relies on XML, appropriate security mechanisms, which are discussed in Chapter 2, have to be applied.

Inspired by the scientific objectives of the SemCrypt research project, the goal of this thesis was to investigate existing general security concepts with special focus on access control, data encryption and key management, which both form the core of SemCrypt security components. The ideas presented in this thesis aim at developing a service-oriented framework based on state-of-the-art Web service and related security technologies in order to provide document-centric database functionality through a trusted environment. It was shown how fine-grained encryption of sensitive data and access control based on semantic concepts can be used to design a collaborative business model, which is secure and still flexible.

Furthermore, a set of service interfaces has been specified, which abstract security and data management functionalities from underlying implementations and can be configured to suit a wide range of different application scenarios as discussed in Chapter 3. Chapter 4 introduces potential architectural settings and illustrates the interactions among entities involved in typical SemCrypt applications. Finally, Chapter 7 discusses concrete implementations, which take advantage of the concepts developed and demonstrate the validity of the framework.

Still, according to [99] there are many open problems, which require a sophisticated research agenda. In Chapter 9, ongoing and future research efforts are briefly presented. These shall improve the current security framework, provide enhanced development and deployment support for SemCrypt applications and are more focused on the incorporation of semantic knowledge.

Chapter 9

Ongoing and Future Work

A lot of research and development energy has been put in the investigation of related technologies so far. One ongoing issue is the planned integration of XAccess policies and JAAS authorization with the XACML standard. Furthermore, XACML could be used to perform access control on system levels other than XML document fragments, such as whole documents, file resources or system components, including service operation invocations [104]. The latter aspect is strongly related to the intended incorporation of the emerging WS-Policy specification. Further studies of Web service security related standards, is a general objective to improve overall system security and state-of-the-art conformance. Due to many overlapping features XrML and SAML integration for rights and identity management is a primary goal as well. With respect to public key management it might be considerable to keep an eye on future XKMS development efforts.

The XCipher encryption instructions supports XML data encryption to the node level, including elements, attributes and node kind tests, such as text content or commentaries. Although this might be sufficient in most cases a support for filter expressions and thus increasing the level of encryption granularity is considered. The idea of XML versioning has been marginally touched, but could not be further investigated both within the SemCrypt project and in the scope of the thesis as well. The framework specified herein has been designed to be capable of supporting document versioning in future extensions and implementations. The XVersion schema presented, suggests a format for storing versioning information in an XML style, but no comparisons or evaluations regarding related work has been made so far. It would be desirable to automate the versioning process by defining schemas and ontologies that lets the application framework discover and adapt to document modifications. Comparable efforts have been made in the area of

WSDL monitoring and evolution management [138].

As discussed in Chapter 2, the SemCrypt database system provides an indexing mechanism for efficient query/update processing of XML document data. However, in cases another DBMS should be deployed, further research in the area of native XML database integration needs to be done. The SemCrypt Database Framework (SCDBF) [29] already provides appropriate abstraction mechanisms, which are subject to current evaluation efforts.

Regarding workflows, there is up to now no way to define access control criteria. The Business Process Execution Language (BPEL) [91], which is the emerging standard for describing workflows and compositions for Web services, leaves all security aspects to the implementation of compliant workflow engines. The approach taken in [62] respectively can be used to specify access control assertions that can be extended by referencing concepts of ontologies expressed in XML. OWL [108] is used to develop ontologies. Referring to Figure 9.1, an ontology to describe the subject of an assertion, an ontology for describing the objects that can be accessed and an ontology for describing the privileges, are developed. On the second layer of the inheritance hierarchy, principal classes that will be found in most applications are described. Subsequent layers define typical domain dependent classes. The advantage of this ontology hierarchy is its extensibility. For instance, a further class **Employee** that is classified as similar to the class **Staff** may be introduced. If one organization uses the class Staff and the other **Employee** the applicability of the same assertion for both companies could be deduced. With XAccess and XCipher descriptions, which are both subject to detailed discussions in Chapter 5, semantic descriptions of security objects, i.e. XML document nodes, as well as information related to key management are provided. Access rights may be recursively composed by other privileges defined for an application scenario; and access control for XML documents may be defined on a fine granular level in either a permissive or more restrictive manner. Using an XAccess descriptor is also suitable for defining access control to Web services, by providing access control rules for WSDL documents. BPEL descriptions may be subject to access control, in case workflow execution should be restricted or ontology information provided by an RDF or OWL document may be protected as well. XAccess is therefore applicable to any XML formatted information. The advantage of XAccess being an XML proposal itself facilitates integration into other XML based standards. Ontology description integration allows the definition of semantic cross-domain information, moving our proposals to an even higher level of abstraction. Although more complex semantic issues are already taken care of, it may

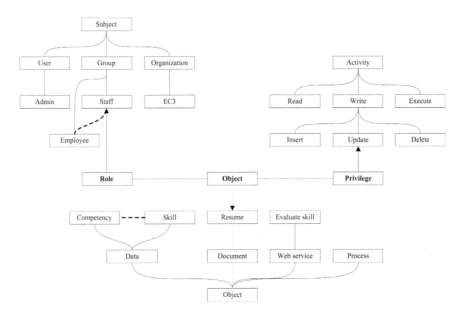

Figure 9.1: Integration of ontologies with authorization assertions

be desirable to refer to other standards, which already address certain problems and that haven not been considered so far. However, implementation of XAccess and XCipher languages are necessary for integrating semantic knowledge into the final concrete application scenario.

Future work could also address the development of application development to evaluate and further improve the suggested SemCrypt technology framework. Initial tool support and IDE integration is available, nevertheless the features provided are kept to a minimalist level to support basic functionalities. With testing performance of selected system components, one important step towards Quality of Service (QoS) evaluation has been made. However, to provide good advices overall application performance is very crucial and valuable information regarding the likeliness of success of the developed security framework. Standard conformance, applicability of the approaches taken as well as usability are strong arguments for convincing future collaborators and potential industry partners to further support the development of the technology in hand.

Part V

Appendices

Appendix A

Schema Definitions

A.1 XAccess.xsd

The following listing is a full representation of the XAccess Schema file. It describes the document structure of XAccess policy files currently implemented in the SemCrypt access control component. A full discussion of XML based authorization is subject to Chapter 5.

Listing A.1: XAccess.xsd

```
    <?xml version="1.0" encoding="UTF-8"?>
    <schema xmlns="http://www.w3.org/2001/XMLSchema"
        targetNamespace="http://semcrypt.ec3.at/xml/access/types"
        xmlns:tns="http://semcrypt.ec3.at/xml/access/types">
5       <element name="XAccess" type="tns:XAccessType"></element>
        <element name="Permissions" type="tns:PermissionsType"></element>
        <element name="Permission" type="tns:DecisionType"></element>
        <element name="PathValue" type="tns:PathValueType"></element>
        <element name="Description" type="tns:DescriptionType"></element>
10      <element name="RoleName" type="tns:RoleNameType"></element>
        <element name="RightName" type="tns:RightNameType"></element>
        <complexType name="XAccessType">
            <sequence>
                <element ref="tns:Namespaces" maxOccurs="1" minOccurs="1" />
15              <element ref="tns:Permissions" maxOccurs="1" minOccurs="0" />
                <element ref="tns:Denials" maxOccurs="1" minOccurs="0" />
            </sequence>
        </complexType>
        <simpleType name="RoleNameType">
20          <restriction base="string" />
        </simpleType>
        <simpleType name="RightNameType">
            <restriction base="string" />
        </simpleType>
25      <complexType name="PermissionsType">
            <sequence>
                <element ref="tns:Permission" maxOccurs="unbounded"
                    minOccurs="1" />
            </sequence>
30      </complexType>
        <simpleType name="PathValueType">
            <restriction base="string" />
        </simpleType>
        <simpleType name="DescriptionType">
35          <restriction base="string" />
        </simpleType>
        <element name="RequiresRoles" type="tns:RequiresRolesType" />
        <complexType name="RequiresRolesType">
            <sequence>
40              <element ref="tns:RoleName" maxOccurs="unbounded"
                    minOccurs="1" />
```

```
        </sequence>
    </complexType>
    <element name="Namespace" type="tns:NamespaceType" />
    <element name="Namespaces" type="tns:NamespacesType" />
    <complexType name="NamespacesType">
        <sequence>
            <element ref="tns:Namespace" maxOccurs="unbounded"
                minOccurs="1" />
        </sequence>
    </complexType>
    <element name="Prefix" type="tns:PrefixType" />
    <element name="URI" type="tns:URIType" />
    <simpleType name="PrefixType">
        <restriction base="string" />
    </simpleType>
    <simpleType name="URIType">
        <restriction base="anyURI" />
    </simpleType>
    <element name="Denials" type="tns:DenialsType" />
    <element name="Denial" type="tns:DecisionType" />
    <complexType name="DenialsType">
        <sequence>
            <element ref="tns:Denial" maxOccurs="unbounded"
                minOccurs="1" />
        </sequence>
    </complexType>
    <complexType name="DecisionType">
        <sequence>
            <element ref="tns:Description" maxOccurs="1" minOccurs="0" />
            <element ref="tns:PathValue" maxOccurs="1" minOccurs="1" />

            <element ref="tns:RoleName" maxOccurs="unbounded"
                minOccurs="1" />
            <element ref="tns:RightName" maxOccurs="unbounded"
                minOccurs="1" />
            <element ref="tns:RequiresRoles" maxOccurs="1"
                minOccurs="0" />
        </sequence>
    </complexType>
    <complexType name="NamespaceType">
        <sequence>
            <element ref="tns:Prefix" />
            <element ref="tns:URI" />
        </sequence>
    </complexType>
</schema>
```

A.2 XCipher.xsd

XCipher documents are currently used by SemCrypt applications to specify
encryption procedures of XML elements identified via XPath. A full description of the ciphering component is subject to Chapter 5.

Listing A.2: XCipher.xsd

```
<?xml version="1.0" encoding="UTF-8"?>
<schema targetNamespace="http://semcrypt.ec3.at/xml/cipher/types"
    xmlns="http://www.w3.org/2001/XMLSchema"
    xmlns:tns="http://semcrypt.ec3.at/xml/cipher/types">
    <element name="XCipher" type="tns:XCipherType"></element>
    <element name="Encryption" type="tns:EncryptionType"></element>
    <element name="Sequence" type="tns:SequenceType"></element>
    <element name="PathValue" type="tns:PathValueType"></element>
    <element name="KeyAlias" type="tns:KeyAliasType"></element>
    <simpleType name="KeyAliasType">
        <restriction base="string"></restriction>
    </simpleType>
    <complexType name="XCipherType">
        <sequence>
            <element ref="tns:Encryption" maxOccurs="unbounded"
                minOccurs="1">
            </element>
        </sequence>
    </complexType>
```

```
20      <complexType name="EncryptionType">
            <sequence>
                <element ref="tns:PathValue" maxOccurs="unbounded" minOccurs="1"></element>
                <element ref="tns:Sequence" maxOccurs="1"
                    minOccurs="1">
25              </element>
            </sequence>
        </complexType>
        <complexType name="SequenceType">
            <sequence>
30              <element ref="tns:KeyAlias" maxOccurs="unbounded"
                    minOccurs="1">
                </element>
            </sequence>
        </complexType>
35      <simpleType name="PathValueType">
            <restriction base="string"></restriction>
        </simpleType>
    </schema>
```

A.3   XVersion.xsd

The following XML Schema provides a structure for storing modification
information about versioned XML documents. XVersion embodies the third
component of SemCrypt XML document handling and is subject to future
enhancements.

Listing A.3: XVersion.xsd

```
    <?xml version="1.0" encoding="UTF-8"?>
    <schema xmlns="http://www.w3.org/2001/XMLSchema"
        targetNamespace="http://semcrypt.ec3.at/xml/version/types"
        xmlns:tns="http://semcrypt.ec3.at/xml/version/types">
5       <element name="XVersion" type="tns:XVersionType"></element>
        <element name="VersionNumber" type="tns:VersionNumberType"></element>
        <element name="Filename" type="tns:FilenameType"></element>
        <element name="CreationInfo" type="tns:CreationInfoType"></element>
        <element name="LastModification"
10          type="tns:LastModificationType">
        </element>
        <element name="Modifications" type="tns:ModificationsType"></element>
        <element name="Author" type="tns:AuthorType"></element>
        <complexType name="LastModificationType">
15          <sequence>
                <element ref="tns:ModificationNumber" minOccurs="1"
                    maxOccurs="1">
                </element>
            </sequence>
20      </complexType>
        <complexType name="AuthorType">
            <sequence>
                <element ref="tns:Firstname" minOccurs="1"></element>
                <element ref="tns:Lastname" minOccurs="1" maxOccurs="1"></element>
25              <element ref="tns:ResponsibleEmail" minOccurs="1"
                    maxOccurs="1">
                </element>
            </sequence>
        </complexType>
30      <complexType name="ModificationsType">
            <sequence>
                <element ref="tns:Modification" minOccurs="1"
                    maxOccurs="unbounded">
                </element>
35          </sequence>
        </complexType>
        <complexType name="CreationInfoType">
            <sequence>
                <element ref="tns:Date" minOccurs="1" maxOccurs="1"></element>
40              <element ref="tns:ResponsibleEmail" minOccurs="1"
                    maxOccurs="1">
                </element>
                <element ref="tns:Purpose" minOccurs="1" maxOccurs="1"></element>
            </sequence>
```

```xml
        </complexType>
        <complexType name="XVersionType">
            <sequence>
                <element ref="tns:VersionNumber" minOccurs="1"
                    maxOccurs="1">
                </element>
                <element ref="tns:Filename" minOccurs="1" maxOccurs="1"></element>
                <element ref="tns:CreationInfo"></element>
                <element ref="tns:LastModification" minOccurs="1"
                    maxOccurs="1">
                </element>
                <element ref="tns:Modifications" minOccurs="1"
                    maxOccurs="1">
                </element>
                <element ref="tns:Authors"></element>
            </sequence>
        </complexType>
        <element name="Date" type="tns:DateType"></element>
        <element name="Purpose" type="tns:PurposeType"></element>
        <element name="ResponsibleEmail"
            type="tns:ResponsibleEmailType">
        </element>
        <element name="Documentation" type="tns:DocumentationType"></element>
        <element name="Modification" type="tns:ModificationType"></element>
        <element name="ModificationNumber"
            type="tns:ModificationNumberType">
        </element>
        <complexType name="ModificationType">
            <sequence>
                <element ref="tns:ModificationNumber" minOccurs="1"
                    maxOccurs="1">
                </element>
                <element ref="tns:Date" minOccurs="1" maxOccurs="1"></element>
                <element ref="tns:ResponsibleEmail" minOccurs="1"
                    maxOccurs="1">
                </element>
                <element ref="tns:Documentation" minOccurs="1"
                    maxOccurs="1">
                </element>
            </sequence>
        </complexType>
        <element name="Firstname" type="tns:FirstnameType"></element>
        <element name="Lastname" type="tns:LastnameType"></element>
        <element name="Authors" type="tns:AuthorsType"></element>
        <complexType name="AuthorsType">
            <sequence>
                <element ref="tns:Author" minOccurs="1"
                    maxOccurs="unbounded">
                </element>
            </sequence>
        </complexType>
        <simpleType name="DateType">
            <restriction base="dateTime"></restriction>
        </simpleType>
        <simpleType name="DocumentationType">
            <restriction base="string"></restriction>
        </simpleType>
        <simpleType name="FilenameType">
            <restriction base="string"></restriction>
        </simpleType>
        <simpleType name="FirstnameType">
            <restriction base="string"></restriction>
        </simpleType>
        <simpleType name="LastnameType">
            <restriction base="string"></restriction>
        </simpleType>
        <simpleType name="ModificationNumberType">
            <restriction base="nonNegativeInteger"></restriction>
        </simpleType>
        <simpleType name="PurposeType">
            <restriction base="string"></restriction>
        </simpleType>
        <simpleType name="ResponsibleEmailType">
            <restriction base="string"></restriction>
        </simpleType>
        <simpleType name="VersionNumberType">
            <restriction base="string"></restriction>
        </simpleType>
</schema>
```

A.4   SemCryptEmployeeInfo.xsd

The following schema represents a document structure compatible with the
SemCrypt storage provider. It allows modeling employees' personal information, assessment results and competencies and serves as reference scenario
for the SemCrypt prototype implementation.

Listing A.4: SemCryptEmployeeInfo.xsd

```
<?xml version="1.0" encoding="UTF-8"?>
<xs:schema xmlns="http://ns.hr-xml.org/2004-08-02"
    xmlns:xs="http://www.w3.org/2001/XMLSchema"
    targetNamespace="http://ns.hr-xml.org/2004-08-02"
5   elementFormDefault="qualified">
    <xs:element name="SemCryptEmployeeInfo">
        <xs:complexType>
            <xs:sequence>
                <xs:element name="PersonInfo"
10                  type="SemCrypt_PersonInfoType" />
                <xs:element name="Competencies"
                    type="SemCrypt_CompetenciesType" minOccurs="0" />
                <xs:element name="JobPositionHistory"
                    type="SemCrypt_JobPositionHistoryType"
15                  minOccurs="0" />
                <xs:element name="Salary" type="SemCrypt_SalaryType"
                    minOccurs="0" />
                <xs:element name="AssessmentResults"
                    type="SemCrypt_AssessmentResultsType"
20                  minOccurs="0" />
            </xs:sequence>
        </xs:complexType>
    </xs:element>
    <xs:complexType name="EntityIdType">
25      <xs:sequence>
            <xs:element name="IdValue" maxOccurs="unbounded">
                <xs:complexType>
                    <xs:simpleContent>
                        <xs:extension base="xs:string">
30                          <xs:attribute name="name" type="xs:string"
                                use="optional" />
                        </xs:extension>
                    </xs:simpleContent>
                </xs:complexType>
35          </xs:element>
        </xs:sequence>
        <xs:attribute name="validFrom" type="xs:date" use="optional" />
        <xs:attribute name="validTo" type="xs:date" use="optional" />
        <xs:attribute name="idOwner" type="xs:string" use="optional" />
40  </xs:complexType>
    <!-- EntityReferenceType-->
    <xs:complexType name="EntityReferenceType">
        <xs:sequence>
            <xs:element name="Id" type="EntityIdType" minOccurs="0" />
45          <xs:element name="Name" type="xs:string" minOccurs="0" />
        </xs:sequence>
    </xs:complexType>
    <xs:complexType name="PersonNameType">
        <xs:sequence>
50          <xs:element name="FormattedName" minOccurs="0"
                maxOccurs="unbounded">
                <xs:complexType>
                    <xs:simpleContent>
                        <xs:extension base="xs:string">
55                          <xs:attribute name="type"
                                type="xs:string" />
                        </xs:extension>
                    </xs:simpleContent>
                </xs:complexType>
60          </xs:element>
            <xs:element name="LegalName" type="xs:string"
                minOccurs="0" />
            <xs:element name="GivenName" type="xs:string"
                minOccurs="0" maxOccurs="unbounded" />
65          <xs:element name="PreferredGivenName" type="xs:string"
                minOccurs="0" />
            <xs:element name="MiddleName" type="xs:string"
                minOccurs="0" />
            <xs:element name="FamilyName" minOccurs="0"
```

```
                    maxOccurs="unbounded">
                <xs:complexType>
                    <xs:simpleContent>
                        <xs:extension base="xs:string">
                            <xs:attribute name="primary"
                                type="xs:string" />
                            <xs:attribute name="prefix"
                                type="xs:string" />
                        </xs:extension>
                    </xs:simpleContent>
                </xs:complexType>
            </xs:element>
            <xs:element name="Affix" minOccurs="0"
                maxOccurs="unbounded">
                <xs:complexType>
                    <xs:simpleContent>
                        <xs:extension base="xs:string">
                            <xs:attribute name="type" type="xs:string"
                                use="required" />
                        </xs:extension>
                    </xs:simpleContent>
                </xs:complexType>
            </xs:element>
        </xs:sequence>
    </xs:complexType>
    <xs:complexType name="PostalAddressType">
        <xs:sequence>
            <xs:element name="CountryCode" type="xs:string" />
            <xs:element name="PostalCode" type="xs:string"
                minOccurs="0" />
            <xs:element name="Region" type="xs:string" minOccurs="0"
                maxOccurs="unbounded" />
            <xs:element name="Municipality" type="xs:string"
                minOccurs="0" />
            <xs:element name="DeliveryAddress" minOccurs="0">
                <xs:complexType>
                    <xs:sequence>
                        <xs:element name="AddressLine"
                            type="xs:string" minOccurs="0"
                            maxOccurs="unbounded" />
                        <xs:element name="StreetName" type="xs:string"
                            minOccurs="0" />
                        <xs:element name="BuildingNumber"
                            type="xs:string" minOccurs="0" />
                        <xs:element name="Unit" type="xs:string"
                            minOccurs="0" />
                        <xs:element name="PostOfficeBox"
                            type="xs:string" minOccurs="0" />
                    </xs:sequence>
                </xs:complexType>
            </xs:element>
            <xs:element name="Recipient" minOccurs="0"
                maxOccurs="unbounded">
                <xs:complexType>
                    <xs:sequence>
                        <xs:element name="PersonName"
                            type="PersonNameType" minOccurs="0" />
                        <xs:element name="AdditionalText"
                            type="xs:string" minOccurs="0"
                            maxOccurs="unbounded" />
                        <xs:element name="Organization"
                            type="xs:string" minOccurs="0" />
                        <xs:element name="OrganizationName"
                            type="xs:string" minOccurs="0" />
                    </xs:sequence>
                </xs:complexType>
            </xs:element>
        </xs:sequence>
        <xs:attribute name="type" type="xs:string" />
    </xs:complexType>
    <xs:complexType name="TelcomNumberType">
        <xs:sequence>
            <xs:element name="FormattedNumber" type="xs:string" />
            <xs:element name="InternationalCountryCode"
                type="xs:string" />
            <xs:element name="NationalNumber" type="xs:string" />
            <xs:element name="AreaCityCode" type="xs:string" />
            <xs:element name="SubscriberNumber" type="xs:string" />
            <xs:element name="Extension" type="xs:string" />
        </xs:sequence>
    </xs:complexType>
    <xs:complexType name="MobileTelcomNumberType">
        <xs:complexContent>
```

```
                  <xs:extension base="TelcomNumberType">
                     <xs:attribute name="smsEnabled" type="xs:boolean"
155                     use="optional" />
                  </xs:extension>
               </xs:complexContent>
            </xs:complexType>
            <xs:complexType name="TelcomNumberListType">
160            <xs:sequence>
                  <xs:element name="Telephone" type="TelcomNumberType"
                     minOccurs="0" />
                  <xs:element name="Mobile" type="MobileTelcomNumberType"
                     minOccurs="0" />
165               <xs:element name="Fax" type="TelcomNumberType"
                     minOccurs="0" />
                  <xs:element name="Pager" type="TelcomNumberType"
                     minOccurs="0" />
                  <xs:element name="TTYTDD" type="TelcomNumberType"
170                  minOccurs="0" />
               </xs:sequence>
            </xs:complexType>
            <xs:complexType name="ContactMethodType">
               <xs:sequence>
175               <xs:element name="Use" type="xs:string" minOccurs="0" />
                  <xs:element name="Location" type="xs:string"
                     minOccurs="0" />
                  <xs:element name="WhenAvailable" type="xs:string"
                     minOccurs="0" />
180               <xs:element name="Telephone" type="TelcomNumberType"
                     minOccurs="0" />
                  <xs:element name="Mobile" type="MobileTelcomNumberType"
                     minOccurs="0" />
                  <xs:element name="Fax" type="TelcomNumberType"
185                  minOccurs="0" />
                  <xs:element name="Pager" type="TelcomNumberType"
                     minOccurs="0" />
                  <xs:element name="TTYTDD" type="TelcomNumberType"
                     minOccurs="0" />
190               <xs:element name="InternetEmailAddress" type="xs:string"
                     minOccurs="0" />
                  <xs:element name="InternetWebAddress" type="xs:string"
                     minOccurs="0" />
                  <xs:element name="PostalAddress" type="PostalAddressType"
195                  minOccurs="0" />
               </xs:sequence>
            </xs:complexType>
            <!-- ==================== flexible dates
                 ============================ -->
200         <xs:complexType name="FlexibleDatesType">
               <xs:sequence>
                  <xs:element name="AnyDate" type="xs:date" />
                  <xs:element name="YearMonth" type="xs:date" />
                  <xs:element name="Year" type="xs:date" />
205               <xs:element name="MonthDay" type="xs:date" />
                  <xs:element name="StringDate" type="xs:string" />
               </xs:sequence>
               <xs:attribute name="dateDescription" type="xs:string" />
            </xs:complexType>
210         <xs:complexType name="UserAreaType">
               <xs:sequence />
               <!-- minOccurs="0" maxOccurs="unbounded" not supported -->
            </xs:complexType>
            <!-- Distribution Guidelines -->
215         <xs:complexType name="DistributionGuidelinesType">
               <xs:sequence>
                  <xs:element name="DistributeTo" type="DistributionType"
                     minOccurs="0" maxOccurs="unbounded" />
                  <xs:element name="DoNotDistributeTo"
220                  type="DistributionType" minOccurs="0"
                     maxOccurs="unbounded" />
               </xs:sequence>
            </xs:complexType>
            <xs:complexType name="DistributionType">
225            <xs:sequence>
                  <xs:element name="Id" type="EntityIdType" minOccurs="0" />
                  <xs:element name="Name" type="xs:string" minOccurs="0" />
               </xs:sequence>
               <xs:attribute name="validFrom" type="xs:date" />
230            <xs:attribute name="validTo" type="xs:date" />
            </xs:complexType>
            <!-- Location Summary -->
            <xs:complexType name="LocationSummaryType">
               <xs:sequence>
235               <xs:element name="CountryCode" type="xs:string"
```

```
                minOccurs="0" />
            <xs:element name="PostalCode" type="xs:string"
                minOccurs="0" />
            <xs:element name="Municipality" type="xs:string"
                minOccurs="0" />
            <xs:element name="Region" type="xs:string" minOccurs="0" />
        </xs:sequence>
    </xs:complexType>
    <!-- Reference Type -->
    <xs:complexType name="ReferenceType">
        <xs:sequence>
            <xs:element name="PersonName" type="PersonNameType"
                minOccurs="0" />
            <xs:element name="PositionTitle" type="xs:string"
                minOccurs="0" />
            <xs:element name="ContactMethod" type="ContactMethodType"
                minOccurs="0" maxOccurs="unbounded" />
            <xs:element name="Comments" type="xs:string"
                minOccurs="0" />
        </xs:sequence>
        <xs:attribute name="type" type="xs:string" />
    </xs:complexType>
    <!-- Acheivement Type -->
    <xs:complexType name="AchievementType">
        <xs:sequence>
            <xs:element name="Date" type="FlexibleDatesType"
                minOccurs="0" />
            <xs:element name="Description" type="xs:string"
                minOccurs="0" />
            <xs:element name="IssuingAuthority" type="xs:string"
                minOccurs="0" />
        </xs:sequence>
    </xs:complexType>
    <!-- Association Type -->
    <xs:complexType name="AssociationType">
        <xs:sequence>
            <xs:element name="Name" type="xs:string" />
            <xs:element name="Id" type="EntityIdType" minOccurs="0" />
            <xs:element name="Link" type="xs:string" minOccurs="0" />
            <xs:element name="StartDate" type="FlexibleDatesType"
                minOccurs="0" />
            <xs:element name="EndDate" type="FlexibleDatesType"
                minOccurs="0" />
            <xs:element name="Role" minOccurs="0"
                maxOccurs="unbounded">
                <xs:complexType>
                    <xs:sequence>
                        <xs:element name="Name" type="xs:string"
                            minOccurs="0" />
                        <xs:element name="Deliverable"
                            type="xs:string" minOccurs="0"
                            maxOccurs="unbounded" />
                        <xs:element name="StartDate"
                            type="FlexibleDatesType" minOccurs="0" />
                        <xs:element name="EndDate"
                            type="FlexibleDatesType" minOccurs="0" />
                        <xs:element name="Comments" type="xs:string"
                            minOccurs="0" />
                    </xs:sequence>
                </xs:complexType>
            </xs:element>
            <xs:element name="Comments" type="xs:string"
                minOccurs="0" />
        </xs:sequence>
    </xs:complexType>
    <!-- Association Types -->
    <!-- SupportingMaterials -->
    <xs:complexType name="StaffingSupportingMaterialsType">
        <xs:sequence>
            <xs:element name="Link" type="xs:string" minOccurs="0" />
            <xs:element name="AttachmentReference" minOccurs="0">
                <xs:complexType>
                    <xs:simpleContent>
                        <xs:extension base="xs:string">
                            <xs:attribute name="context"
                                type="xs:string" />
                            <xs:attribute name="mimeType"
                                type="xs:string" />
                        </xs:extension>
                    </xs:simpleContent>
                </xs:complexType>
            </xs:element>
            <xs:element name="Description" type="xs:string"
```

```
                          minOccurs="0" />
320               </xs:sequence>
          </xs:complexType>
          <!-- Language -->
          <!-- Job Level Info   -->
          <xs:complexType name="JobLevelInfoType">
325           <xs:sequence>
                  <xs:element name="JobPlan" type="xs:string" minOccurs="0" />
                  <xs:element name="JobGrade" type="xs:string"
                      minOccurs="0" />
                  <xs:element name="JobStep" type="xs:string" minOccurs="0" />
330               <xs:element name="Comments" type="xs:string"
                      minOccurs="0" />
              </xs:sequence>
          </xs:complexType>
          <!-- Occupational Taxonomy Types   -->
335       <xs:complexType name="OccupationalCategoryType">
              <xs:sequence>
                  <xs:element name="TaxonomyName" minOccurs="0">
                      <xs:complexType>
                          <xs:simpleContent>
340                           <xs:extension base="xs:string">
                                  <xs:attribute name="version"
                                      type="xs:string" use="optional" />
                              </xs:extension>
                          </xs:simpleContent>
345                       </xs:complexType>
                  </xs:element>
                  <xs:element name="CategoryCode" type="xs:string"
                      minOccurs="0" />
                  <xs:element name="CategoryDescription" type="xs:string"
350                   minOccurs="0" />
                  <xs:element name="Comments" type="xs:string"
                      minOccurs="0" />
                  <xs:element name="JobCategory"
                      type="OccupationalCategoryType" minOccurs="0" />
355           </xs:sequence>
          </xs:complexType>
          <!-- Position Posting Type -->
          <xs:complexType name="PositionPostingsType">
              <xs:sequence>
360               <xs:element name="PositionPosting" minOccurs="0"
                      maxOccurs="unbounded">
                      <xs:complexType>
                          <xs:sequence>
                              <xs:element name="Id" type="EntityIdType"
365                               minOccurs="0" />
                              <xs:element name="Title" type="xs:string"
                                  minOccurs="0" />
                              <xs:element name="Link" type="xs:string"
                                  minOccurs="0" />
370                       </xs:sequence>
                      </xs:complexType>
                  </xs:element>
              </xs:sequence>
          </xs:complexType>
375       <!-- Supplier Type -->
          <xs:complexType name="SupplierType">
              <xs:sequence>
                  <xs:element name="SupplierId" type="EntityIdType"
                      minOccurs="0" />
380               <xs:element name="EntityName" type="xs:string"
                      minOccurs="0" />
                  <xs:element name="ContactMethod" type="ContactMethodType"
                      minOccurs="0" maxOccurs="unbounded" />
              </xs:sequence>
385           <xs:attribute name="relationship" type="xs:string"
                  use="optional" />
          </xs:complexType>
          <!-- RecordInfoType -->
          <xs:complexType name="RecordInfoType">
390           <xs:sequence>
                  <xs:element name="Id" type="EntityIdType" minOccurs="0"
                      maxOccurs="unbounded" />
                  <!-- <xs:element name="Status" minOccurs="0">
                      <xs:complexType>
395                       <xs:simpleContent><xs:extension>
                          <xs:attribute name="validFrom"
                              type="xs:date" />
                          <xs:attribute name="validTo"
                              type="xs:date" />
400                       </xs:extension>
                      </xs:simpleContent>
```

```
            </xs:complexType>
            </xs:element> -->
        </xs:sequence>
    </xs:complexType>
    <!-- PositionDateInfo type -->
    <xs:complexType name="PositionDateInfoType">
        <xs:sequence>
            <xs:element name="StartAsSoonAsPossible" type="xs:boolean"
                minOccurs="0" />
            <xs:element name="StartDate" type="xs:date" minOccurs="0" />
            <xs:element name="ExpectedEndDate" type="xs:date"
                minOccurs="0" />
            <xs:element name="MaximumStartDate" type="xs:date"
                minOccurs="0" />
            <xs:element name="MaximumEndDate" type="xs:date"
                minOccurs="0" />
        </xs:sequence>
    </xs:complexType>
    <xs:complexType name="PersonDescriptorsType">
        <xs:sequence>
            <xs:element name="LegalIdentifiers"
                type="LegalIdentifiersType" minOccurs="0" />
            <xs:element name="DemographicDescriptors"
                type="DemographicDescriptorsType" minOccurs="0" />
            <xs:element name="BiologicalDescriptors"
                type="BiologicalDescriptorsType" minOccurs="0" />
            <xs:element name="SupportingMaterials"
                type="StaffingSupportingMaterialsType" minOccurs="0"
                maxOccurs="unbounded" />
            <xs:element name="OtherDescriptors"
                type="OtherDescriptorsType" minOccurs="0"
                maxOccurs="unbounded" />
            <xs:element name="UserArea" type="UserAreaType"
                minOccurs="0" />
        </xs:sequence>
    </xs:complexType>
    <xs:complexType name="LegalIdentifiersType">
        <xs:sequence>
            <xs:element name="PersonLegalId" type="PersonLegalIdType"
                minOccurs="0" maxOccurs="unbounded" />
            <xs:element name="MilitaryStatus" minOccurs="0"
                maxOccurs="unbounded">
                <xs:complexType>
                    <xs:simpleContent>
                        <xs:extension base="xs:string">
                            <xs:attribute name="type"
                                type="xs:string" />
                        </xs:extension>
                    </xs:simpleContent>
                </xs:complexType>
            </xs:element>
            <xs:element name="VisaStatus" minOccurs="0"
                maxOccurs="unbounded">
                <xs:complexType>
                    <xs:simpleContent>
                        <xs:extension base="xs:string">
                            <xs:attribute name="countryCode"
                                type="xs:string" />
                            <xs:attribute name="validFrom"
                                type="xs:date" />
                            <xs:attribute name="validTo"
                                type="xs:date" />
                        </xs:extension>
                    </xs:simpleContent>
                </xs:complexType>
            </xs:element>
            <xs:element name="Citizenship" type="xs:string"
                minOccurs="0" maxOccurs="unbounded" />
            <xs:element name="Residency" type="xs:string"
                minOccurs="0" maxOccurs="unbounded" />
            <xs:element name="UserArea" type="UserAreaType"
                minOccurs="0" />
        </xs:sequence>
    </xs:complexType>
    <xs:complexType name="DemographicDescriptorsType">
        <xs:sequence>
            <xs:element name="Race" type="xs:string" minOccurs="0"
                maxOccurs="unbounded" />
            <xs:element name="Ethnicity" type="xs:string"
                minOccurs="0" maxOccurs="unbounded" />
            <xs:element name="Nationality" type="xs:string"
                minOccurs="0" maxOccurs="unbounded" />
            <xs:element name="PrimaryLanguage" type="xs:string"
```

```
485                      minOccurs="0" maxOccurs="unbounded" />
                     <xs:element name="BirthPlace" type="xs:string"
                         minOccurs="0" />
                     <xs:element name="Religion" type="xs:string"
                         minOccurs="0" />
490                  <xs:element name="MaritalStatus" type="xs:string"
                         minOccurs="0" />
                     <xs:element name="ChildrenInfo" minOccurs="0">
                         <xs:complexType>
                             <xs:sequence>
495                              <xs:element name="NumberOfChildren"
                                     type="xs:integer" minOccurs="0" />
                                 <xs:element name="Comments" type="xs:string"
                                     minOccurs="0" />
                             </xs:sequence>
500                      </xs:complexType>
                     </xs:element>
                     <xs:element name="UserArea" type="UserAreaType"
                         minOccurs="0" />
                 </xs:sequence>
505          </xs:complexType>
         <xs:complexType name="BiologicalDescriptorsType">
             <xs:sequence>
                 <xs:element name="DateOfBirth" type="xs:date"
                     minOccurs="0" />
510              <xs:element name="MonthDayOfBirth" type="xs:date"
                     minOccurs="0" />
                 <xs:element name="Age" type="xs:integer" minOccurs="0" />
                 <xs:element name="GenderCode" type="xs:integer"
                     minOccurs="0" />
515              <xs:element name="EyeColor" type="xs:string"
                     minOccurs="0" />
                 <xs:element name="HairColor" type="xs:string"
                     minOccurs="0" />
                 <xs:element name="Height" minOccurs="0">
520                  <xs:complexType>
                         <xs:simpleContent>
                             <xs:extension base="xs:string">
                                 <xs:attribute name="unitOfMeasure"
                                     type="xs:string" />
525                          </xs:extension>
                         </xs:simpleContent>
                     </xs:complexType>
                 </xs:element>
                 <xs:element name="Weight" minOccurs="0">
530                  <xs:complexType>
                         <xs:simpleContent>
                             <xs:extension base="xs:string">
                                 <xs:attribute name="unitOfMeasure"
                                     type="xs:string" />
535                          </xs:extension>
                         </xs:simpleContent>
                     </xs:complexType>
                 </xs:element>
                 <xs:element name="IdentifyingMarks" type="xs:string"
540                  minOccurs="0" maxOccurs="unbounded" />
                 <xs:element name="DisabilityInfo"
                     type="DisabilityInfoType" minOccurs="0"
                     maxOccurs="unbounded" />
                 <xs:element name="UserArea" type="UserAreaType"
545                  minOccurs="0" />
             </xs:sequence>
         </xs:complexType>
         <xs:complexType name="OtherDescriptorsType">
             <xs:sequence>
550              <xs:element name="Name" type="xs:string" minOccurs="0" />
                 <xs:element name="Applicable" type="xs:boolean"
                     minOccurs="0" />
                 <xs:element name="Value" type="xs:string" minOccurs="0" />
                 <xs:element name="List" minOccurs="0">
555                  <xs:complexType>
                         <xs:sequence>
                             <xs:element name="Item" type="xs:string"
                                 maxOccurs="unbounded" />
                         </xs:sequence>
560                  </xs:complexType>
                 </xs:element>
             </xs:sequence>
         </xs:complexType>
         <xs:complexType name="PersonLegalIdType">
565          <xs:complexContent>
                 <xs:extension base="EntityIdType">
                     <xs:attribute name="countryCode" type="xs:string" />
```

```
                <xs:attribute name="jurisdiction" type="xs:string" />
                <xs:attribute name="issuingRegion" type="xs:string" />
                <xs:attribute name="documentType" type="xs:string" />
                <xs:attribute name="idSource" type="xs:string" />
            </xs:extension>
        </xs:complexContent>
    </xs:complexType>
    <xs:complexType name="DisabilityInfoType">
        <xs:sequence>
            <xs:element name="LevelOfDisability" type="xs:string"
                minOccurs="0" />
            <xs:element name="Percentage" type="xs:integer"
                minOccurs="0" />
            <xs:element name="Type" type="xs:string" minOccurs="0" />
            <xs:element name="AccommodationsNeeded" type="xs:boolean"
                minOccurs="0" />
        </xs:sequence>
    </xs:complexType>
    <xs:complexType name="JobIdentifierType">
        <xs:sequence>
            <xs:element name="Id" type="xs:string" />
            <xs:element name="Domain" minOccurs="0">
                <xs:complexType>
                    <xs:sequence>
                        <xs:element name="IdIssuer" type="xs:string"
                            minOccurs="0" />
                        <xs:element name="IdType" type="xs:string"
                            minOccurs="0" />
                    </xs:sequence>
                </xs:complexType>
            </xs:element>
        </xs:sequence>
    </xs:complexType>
    <xs:complexType name="PositionIdentifierType">
        <xs:sequence>
            <xs:element name="Id" type="xs:string" />
            <xs:element name="Domain" minOccurs="0">
                <xs:complexType>
                    <xs:sequence>
                        <xs:element name="IdIssuer" type="xs:string"
                            minOccurs="0" />
                        <xs:element name="IdType" type="xs:string"
                            minOccurs="0" />
                    </xs:sequence>
                </xs:complexType>
            </xs:element>
        </xs:sequence>
    </xs:complexType>
    <xs:complexType name="CalculatedValueFixedType">
        <xs:sequence>
            <xs:element name="Value" type="xs:double" minOccurs="0" />
        </xs:sequence>
    </xs:complexType>
    <!-- Calculated Argument Types -->
    <xs:complexType name="ArgumentType">
        <xs:complexContent>
            <xs:extension base="CalculatedArgumentType">
                <xs:attribute name="index" type="xs:integer" />
            </xs:extension>
        </xs:complexContent>
    </xs:complexType>
    <xs:complexType name="CalculatedArgumentType">
        <xs:sequence>
            <xs:element name="ArgumentValue" type="ArgumentValueType" />
            <xs:element name="ArgumentVariableName"
                type="ArgumentVariableType" />
        </xs:sequence>
    </xs:complexType>
    <xs:complexType name="ArgumentValueType">
        <xs:simpleContent>
            <xs:extension base="xs:string">
                <xs:attribute name="name" type="xs:string" />
                <xs:attribute name="multiplier" type="xs:double" />
            </xs:extension>
        </xs:simpleContent>
    </xs:complexType>
    <xs:complexType name="ArgumentVariableType">
        <xs:simpleContent>
            <xs:extension base="xs:string">
                <xs:attribute name="name" type="xs:string" />
                <xs:attribute name="multiplier" type="xs:double" />
            </xs:extension>
        </xs:simpleContent>
```

```
          </xs:complexType>
          <!-- Basis Type -->
          <xs:complexType name="BasisType">
              <xs:sequence>
655               <xs:element name="BasisValue" type="ArgumentValueType" />
                  <xs:element name="BasisVariableName"
                      type="ArgumentVariableType" />
              </xs:sequence>
          </xs:complexType>
660       <!-- the Calculated Value Percent   -->
          <xs:complexType name="CalculatedValuePercentType">
              <xs:sequence>
                  <xs:element name="Basis" type="BasisType" />
                  <xs:element name="Percentage" type="xs:double" />
665           </xs:sequence>
          </xs:complexType>
          <!-- the Calculated Value Linear  -->
          <xs:complexType name="CalculatedValueLinearType">
              <xs:sequence>
670               <xs:element name="YIntercept" type="xs:double"
                      minOccurs="0" />
                  <xs:element name="Argument" type="ArgumentType"
                      maxOccurs="unbounded" />
              </xs:sequence>
675       </xs:complexType>
          <!-- the Calculated Lookup Table -->
          <xs:complexType name="CalculatedLookupTableEntryType">
              <xs:sequence>
                  <xs:element name="Key" type="xs:string" />
680               <xs:element name="Value" type="ArgumentValueType" />
              </xs:sequence>
          </xs:complexType>
          <xs:complexType name="CalculatedLookupTableType">
              <xs:sequence>
685               <xs:element name="LookupTableId" type="EntityIdType"
                      minOccurs="0" />
                  <xs:element name="LookupTableName" type="xs:string"
                      minOccurs="0" />
                  <xs:element name="LookupTableEntry"
690                   type="CalculatedLookupTableEntryType"
                      maxOccurs="unbounded" />
              </xs:sequence>
          </xs:complexType>
          <!-- the Lookup Table -->
695       <xs:complexType name="CalculatedValueLookupType">
              <xs:sequence>
                  <xs:element name="LookupTableId" type="EntityIdType"
                      minOccurs="0" />
                  <xs:element name="LookupTable"
700                   type="CalculatedLookupTableType" minOccurs="0" />
                  <xs:element name="LookupKey" type="xs:string" />
              </xs:sequence>
              <xs:attribute name="notFoundValue" type="xs:double" />
          </xs:complexType>
705       <!-- the Calculated Schedule Interval Type -->
          <!-- the Calculated Schedule double Interval  -->
          <xs:complexType name="CalculatedScheduleIntervalValueType">
              <xs:simpleContent>
                  <xs:extension base="xs:double">
710                   <xs:attribute name="inclusive" type="xs:boolean" />
                  </xs:extension>
              </xs:simpleContent>
          </xs:complexType>
          <xs:complexType name="CalculatedScheduleTableEntryType">
715           <xs:sequence>
                  <xs:element name="IntervalLow"
                      type="CalculatedScheduleIntervalValueType"
                      minOccurs="0" />
                  <xs:element name="IntervalHigh"
720                   type="CalculatedScheduleIntervalValueType"
                      minOccurs="0" />
                  <xs:element name="Fixed" type="CalculatedValueFixedType" />
                  <xs:element name="Percent"
                      type="CalculatedValuePercentType" />
725               <xs:element name="Linear"
                      type="CalculatedValueLinearType" />
                  <xs:element name="Lookup"
                      type="CalculatedValueLookupType" />
              </xs:sequence>
730       </xs:complexType>
          <xs:complexType name="CalculatedScheduleTableType">
              <xs:sequence>
                  <xs:element name="ScheduleTableId" type="EntityIdType"
```

```
                    minOccurs="0" />
            <xs:element name="ScheduleTableName" type="xs:string"
                    minOccurs="0" />
            <xs:element name="ScheduleTableEntry"
                    type="CalculatedScheduleTableEntryType"
                    maxOccurs="unbounded" />
        </xs:sequence>
    </xs:complexType>
    <!-- the Schedule Table -->
    <xs:complexType name="CalculatedValueScheduleType">
        <xs:sequence>
            <xs:element name="ScheduleTableId" type="EntityIdType"
                    minOccurs="0" />
            <xs:element name="ScheduleTable"
                    type="CalculatedScheduleTableType" minOccurs="0" />
            <xs:element name="ScheduleValue"
                    type="CalculatedScheduleIntervalValueType" />
        </xs:sequence>
        <xs:attribute name="notFoundValue" type="xs:double" />
    </xs:complexType>
    <!-- the Calculated Value Extended   -->
    <xs:complexType name="CalculatedValueExtendedType">
        <xs:sequence>
            <xs:element name="Function" type="xs:string" />
            <xs:element name="Argument" type="ArgumentType"
                    minOccurs="0" maxOccurs="unbounded" />
        </xs:sequence>
    </xs:complexType>
    <!-- the Calculated value -->
    <xs:complexType name="CalculatedValueType">
        <xs:sequence>
            <xs:element name="Fixed" type="CalculatedValueFixedType" />
            <xs:element name="Percent"
                    type="CalculatedValuePercentType" />
            <xs:element name="Linear"
                    type="CalculatedValueLinearType" />
            <xs:element name="Lookup"
                    type="CalculatedValueLookupType" />
            <xs:element name="Schedule"
                    type="CalculatedValueScheduleType" />
            <xs:element name="Extended"
                    type="CalculatedValueExtendedType" />
        </xs:sequence>
    </xs:complexType>
    <!-- infinity types -->
    <!-- reusable elements -->
    <!-- the Deduction -->
    <xs:complexType name="DeductionType">
        <xs:sequence>
            <xs:element name="ApplyToCompensation" type="xs:string"
                    minOccurs="0" maxOccurs="unbounded" />
            <xs:element name="DeductionPlan" type="DeductionPlanType" />
            <xs:element name="DeductionStartDate" type="xs:date" />
            <xs:element name="DeductionOrder" type="xs:integer"
                    minOccurs="0" />
            <xs:element name="DeductionEndDate" type="xs:date"
                    minOccurs="0" />
            <xs:element name="DeductionFrequency" type="xs:string"
                    minOccurs="0" />
            <xs:element name="DeductionCaseNumber" type="EntityIdType"
                    minOccurs="0" />
            <xs:element name="DeductionOptionalDate" type="xs:string"
                    minOccurs="0" maxOccurs="unbounded" />
            <xs:element name="DeductionPreTaxFlag" type="xs:boolean"
                    minOccurs="0" />
        </xs:sequence>
        <xs:attribute name="reportOnly" type="xs:boolean" />
    </xs:complexType>
    <!-- the Deduction Plan   -->
    <xs:complexType name="DeductionPlanType">
        <xs:sequence>
            <xs:element name="DeductionCategory" type="xs:string" />
            <xs:element name="DeductionPlanName" type="xs:string"
                    minOccurs="0" />
            <xs:element name="DeductionPlanId" type="EntityIdType"
                    minOccurs="0" />
        </xs:sequence>
    </xs:complexType>
    <!-- the Benefit -->
    <xs:complexType name="BenefitType">
        <xs:sequence>
            <xs:element name="ApplyToCompensation" type="xs:string"
                    minOccurs="0" maxOccurs="unbounded" />
```

```
                    <xs:element name="BenefitPlan" type="BenefitPlanType" />
                    <xs:element name="BenefitStartDate" type="xs:date" />
                    <xs:element name="BenefitEndDate" type="xs:date"
820                      minOccurs="0" />
                    <xs:element name="BenefitFrequency" type="xs:string"
                       minOccurs="0" maxOccurs="unbounded" />
                    <xs:element name="BenefitCaseNumber" type="xs:string"
                       minOccurs="0" />
825                 <xs:element name="BenefitOptionalDate" type="xs:string"
                       minOccurs="0" maxOccurs="unbounded" />
             </xs:sequence>
             <xs:attribute name="benefitTaxTreatment" type="xs:string" />
             <xs:attribute name="reportOnly" type="xs:boolean" />
830     </xs:complexType>
        <!-- the Benefit Plan  -->
        <xs:complexType name="BenefitPlanType">
             <xs:sequence>
                    <xs:element name="BenefitCategory" type="xs:string" />
835                 <xs:element name="BenefitPlanName" type="xs:string"
                       minOccurs="0" />
                    <xs:element name="BenefitPlanId" type="EntityIdType"
                       minOccurs="0" />
             </xs:sequence>
840     </xs:complexType>
        <!-- the Amount -->
        <xs:complexType name="PayrollAmountType">
             <xs:sequence>
                    <xs:element name="AmountStartDate" type="xs:date" />
845                 <xs:element name="AmountEndDate" type="xs:date"
                       minOccurs="0" />
                    <xs:element name="AmountValue" type="CalculatedValueType" />
                    <xs:element name="AmountLimit" type="PayrollLimitType"
                       minOccurs="0" maxOccurs="unbounded" />
850         </xs:sequence>
             <xs:attribute name="duration" type="xs:string" />
             <xs:attribute name="count" type="xs:integer" />
             <xs:attribute name="currency" type="xs:string" />
        </xs:complexType>
855     <!-- Limit -->
        <xs:complexType name="PayrollLimitType">
             <xs:sequence>
                    <xs:element name="LimitMaxValue" type="xs:double" />
                    <xs:element name="LimitMinValue" type="xs:double" />
860         </xs:sequence>
             <xs:attribute name="type" type="xs:string" />
        </xs:complexType>
        <!-- payroll date type
             <xs:complexType name="Payrollxs:date">
865         <xs:simpleContent>
             <xs:extension base="xs:date">
             <xs:attribute name="type" type="PayrollOtherxs:date" />
             </xs:extension>
             </xs:simpleContent>
870         </xs:complexType> -->
        <xs:complexType name="AssessmentStatusType">
             <xs:sequence>
                    <xs:element name="Status" type="xs:string" />
                    <xs:element name="Details" type="xs:string" minOccurs="0" />
875                 <xs:element name="StatusDate" type="xs:date" />
             </xs:sequence>
        </xs:complexType>
        <xs:complexType name="AssessmentRequestType">
             <xs:sequence>
880                 <xs:element name="ClientId" type="EntityIdType" />
                    <xs:element name="ProviderId" type="EntityIdType"
                       minOccurs="0" />
                    <xs:element name="ClientOrderId" type="EntityIdType" />
                    <xs:element name="ReceiptId" type="EntityIdType"
885                    minOccurs="0" />
                    <xs:element name="UserArea" type="UserAreaType"
                       minOccurs="0" />
             </xs:sequence>
        </xs:complexType>
890     <xs:complexType name="AssessmentResultType">
             <xs:sequence>
                    <xs:element name="ClientId" type="EntityIdType" />
                    <xs:element name="ProviderId" type="EntityIdType"
                       minOccurs="0" />
895                 <xs:element name="ClientOrderId" type="EntityIdType" />
                    <xs:element name="ReceiptId" type="EntityIdType"
                       minOccurs="0" />
                    <xs:element name="Results" minOccurs="0"
                       maxOccurs="unbounded">
```

```xml
            <xs:complexType>
                <xs:sequence>
                    <xs:element name="Profile" type="xs:string"
                        minOccurs="0" />
                    <xs:element name="OverallResult"
                        type="AssessmentResultsType"
                        minOccurs="0" />
                    <xs:element name="DetailResult"
                        type="AssessmentResultsType" minOccurs="0"
                        maxOccurs="unbounded" />
                </xs:sequence>
            </xs:complexType>
        </xs:element>
        <xs:element name="AssessmentStatus"
            type="AssessmentStatusType" />
        <xs:element name="UserArea" type="UserAreaType"
            minOccurs="0" />
    </xs:sequence>
    <!-- <xs:attribute ref="xml:lang"/>-->
</xs:complexType>
<xs:complexType name="AssessmentResultsType">
    <xs:sequence>
        <xs:element name="ScoreId" type="EntityIdType"
            minOccurs="0" />
        <xs:element name="Passed" type="xs:boolean" minOccurs="0" />
        <xs:element name="Description" type="xs:string"
            minOccurs="0" />
        <xs:element name="Score" minOccurs="0"
            maxOccurs="unbounded">
            <xs:complexType>
                <xs:simpleContent>
                    <xs:extension base="xs:double">
                        <xs:attribute name="type" type="xs:string"
                            use="required" />
                    </xs:extension>
                </xs:simpleContent>
            </xs:complexType>
        </xs:element>
        <xs:element name="Scale" type="xs:string" minOccurs="0" />
        <xs:element name="Band" type="xs:string" minOccurs="0" />
        <xs:element name="Comments" type="xs:string"
            minOccurs="0" />
    </xs:sequence>
</xs:complexType>
<xs:complexType name="SemCrypt_PersonInfoType">
    <xs:sequence>
        <xs:element name="PersonId" type="EntityIdType" />
        <xs:element name="PersonName" type="PersonNameType" />
        <xs:element name="ContactMethod" type="ContactMethodType"
            minOccurs="0" maxOccurs="unbounded" />
        <xs:element name="PersonDescriptors"
            type="PersonDescriptorsType" minOccurs="0"
            maxOccurs="unbounded" />
    </xs:sequence>
</xs:complexType>
<xs:complexType name="SemCrypt_CompetenciesType">
    <xs:sequence>
        <xs:element name="Competency" maxOccurs="unbounded">
            <xs:complexType>
                <xs:sequence>
                    <xs:element name="CompetencyId"
                        minOccurs="0">
                        <xs:complexType>
                            <xs:attribute name="id"
                                type="xs:string" use="required" />
                            <xs:attribute name="idOwner"
                                type="xs:string" />
                            <xs:attribute name="description"
                                type="xs:string" />
                        </xs:complexType>
                    </xs:element>
                    <xs:element name="TaxonomyId" minOccurs="0"
                        maxOccurs="unbounded">
                        <xs:complexType>
                            <xs:attribute name="id"
                                type="xs:string" use="required" />
                            <xs:attribute name="idOwner"
                                type="xs:string" />
                            <xs:attribute name="description"
                                type="xs:string" />
                        </xs:complexType>
                    </xs:element>
                    <xs:element name="CompetencyEvidence"
```

```
                              minOccurs="0" maxOccurs="unbounded">
                              <xs:complexType>
985                             <xs:sequence>
                                  <xs:element name="EvidenceId"
                                    minOccurs="0">
                                    <xs:complexType>
                                      <xs:attribute name="id"
990                                     type="xs:string"
                                        use="required" />
                                      <xs:attribute
                                        name="idOwner"
                                        type="xs:string" />
995                                   <xs:attribute
                                        name="description"
                                        type="xs:string" />
                                    </xs:complexType>
                                  </xs:element>
1000                              <xs:element name="NumericValue">
                                    <xs:complexType>
                                      <xs:simpleContent>
                                        <xs:extension
                                          base="xs:double">
1005                                      <xs:attribute
                                            name="minValue"
                                            type="xs:double" />
                                          <xs:attribute
                                            name="maxValue"
1010                                        type="xs:double" />
                                          <xs:attribute
                                            name="description"
                                            type="xs:string" />
                                        </xs:extension>
1015                                  </xs:simpleContent>
                                    </xs:complexType>
                                  </xs:element>
                                  <xs:element name="StringValue">
                                    <xs:complexType>
1020                                  <xs:simpleContent>
                                        <xs:extension
                                          base="xs:string">
                                          <xs:attribute
                                            name="minValue"
1025                                        type="xs:string" />
                                          <xs:attribute
                                            name="maxValue"
                                            type="xs:string" />
                                          <xs:attribute
1030                                        name="description"
                                            type="xs:string" />
                                        </xs:extension>
                                      </xs:simpleContent>
                                    </xs:complexType>
1035                              </xs:element>
                                  <xs:element
                                    name="SupportingInformation"
                                    type="xs:string" minOccurs="0"
                                    maxOccurs="unbounded" />
1040                            </xs:sequence>
                                <xs:attribute name="dateOfIncident"
                                  type="xs:date" />
                                <xs:attribute name="name"
                                  type="xs:string" />
1045                            <xs:attribute name="typeDescription"
                                  type="xs:string" />
                                <xs:attribute name="expirationDate"
                                  type="xs:date" />
                                <xs:attribute name="typeId"
1050                              type="xs:string" />
                                <xs:attribute name="required"
                                  type="xs:boolean" />
                                <xs:attribute name="lastUsed"
                                  type="xs:date" />
1055                          </xs:complexType>
                            </xs:element>
                            <xs:element name="CompetencyWeight"
                              minOccurs="0" maxOccurs="unbounded">
                              <xs:complexType>
1060                            <xs:sequence>
                                  <xs:element name="NumericValue">
                                    <xs:complexType>
                                      <xs:simpleContent>
                                        <xs:extension
1065                                      base="xs:double">
```

```
                                      <xs:attribute
                                          name="minValue"
                                          type="xs:double" />
                                      <xs:attribute
                                          name="maxValue"
                                          type="xs:double" />
                                      <xs:attribute
                                          name="description"
                                          type="xs:string" />
                                  </xs:extension>
                              </xs:simpleContent>
                          </xs:complexType>
                      </xs:element>
                      <xs:element name="StringValue">
                          <xs:complexType>
                              <xs:simpleContent>
                                  <xs:extension
                                      base="xs:string">
                                      <xs:attribute
                                          name="minValue"
                                          type="xs:string" />
                                      <xs:attribute
                                          name="maxValue"
                                          type="xs:string" />
                                      <xs:attribute
                                          name="description"
                                          type="xs:string" />
                                  </xs:extension>
                              </xs:simpleContent>
                          </xs:complexType>
                      </xs:element>
                      <xs:element
                          name="SupportingInformation"
                          type="xs:string" minOccurs="0"
                          maxOccurs="unbounded" />
                  </xs:sequence>
                  <xs:attribute name="type"
                      type="xs:string" />
              </xs:complexType>
          </xs:element>
          <xs:element name="UserArea"
              type="UserAreaType" minOccurs="0" />
      </xs:sequence>
      <xs:attribute name="name" type="xs:string" />
      <xs:attribute name="description" type="xs:string" />
      <xs:attribute name="required" type="xs:boolean" />
  </xs:complexType>
  </xs:element>
  </xs:sequence>
</xs:complexType>
<xs:complexType name="SemCrypt_JobPositionHistoryType">
  <xs:sequence>
      <xs:element name="JobHeader" minOccurs="0"
          maxOccurs="unbounded">
          <xs:complexType>
              <xs:sequence>
                  <xs:element name="JobId"
                      type="JobIdentifierType" />
                  <xs:element name="JobTitle" type="xs:string"
                      minOccurs="0" />
                  <xs:element name="JobDescription"
                      type="xs:string" minOccurs="0" />
                  <xs:element name="JobStatus" minOccurs="0">
                      <xs:complexType>
                          <xs:sequence>
                              <xs:element name="Code"
                                  type="xs:string"
                                  minOccurs="0" />
                              <xs:element name="Description"
                                  type="xs:string"
                                  minOccurs="0" />
                          </xs:sequence>
                      </xs:complexType>
                  </xs:element>
                  <xs:element name="JobLevel" minOccurs="0"
                      maxOccurs="unbounded">
                      <xs:complexType>
                          <xs:sequence>
                              <xs:element name="Code"
                                  type="xs:string"
                                  minOccurs="0" />
                              <xs:element name="Description"
                                  type="xs:string"
```

```
                                        minOccurs="0" />
1150                            </xs:sequence>
                               <xs:attribute name="type"
                                        type="xs:string" />
                            </xs:complexType>
                        </xs:element>
1155                    <xs:element name="JobCategory" minOccurs="0"
                            maxOccurs="unbounded">
                            <xs:complexType>
                                <xs:sequence>
                                    <xs:element name="Code"
1160                                    type="xs:string"
                                        minOccurs="0" />
                                    <xs:element name="Description"
                                        type="xs:string"
                                        minOccurs="0" />
1165                            </xs:sequence>
                               <xs:attribute name="type"
                                        type="xs:string" />
                            </xs:complexType>
                        </xs:element>
1170                </xs:sequence>
                    <xs:attribute name="validFrom" type="xs:date" />
                    <xs:attribute name="validTo" type="xs:date" />
                </xs:complexType>
            </xs:element>
1175        <xs:element name="PositionHeader" minOccurs="0"
                maxOccurs="unbounded">
                <xs:complexType>
                    <xs:sequence>
                        <xs:element name="PositionId"
1180                        type="PositionIdentifierType" />
                        <xs:element name="JobId"
                            type="PositionIdentifierType"
                            minOccurs="0" />
                        <xs:element name="PositionTitle"
1185                        type="xs:string" minOccurs="0" />
                        <xs:element name="PositionType"
                            type="xs:string" minOccurs="0" />
                        <xs:element name="PositionDescription"
                            type="xs:string" />
1190                    <xs:element name="PositionStatus">
                            <xs:complexType>
                                <xs:sequence>
                                    <xs:element name="Code"
                                        type="xs:string"
1195                                    minOccurs="0" />
                                    <xs:element name="Description"
                                        type="xs:string"
                                        minOccurs="0" />
                                </xs:sequence>
1200                        </xs:complexType>
                        </xs:element>
                        <xs:element name="ReportToPositionId"
                            type="PositionIdentifierType" />
                        <xs:element name="SpecialInstructions"
1205                        type="xs:string" />
                        <xs:element name="TypeOfHours"
                            type="xs:string" minOccurs="0" />
                        <xs:element name="Quantity" type="xs:double"
                            minOccurs="0" />
1210                    <xs:element name="RequestedPerson"
                            minOccurs="0" maxOccurs="unbounded">
                            <xs:complexType>
                                <xs:sequence>
                                    <xs:element name="PersonName"
1215                                    type="PersonNameType"
                                        minOccurs="0" />
                                    <xs:element name="PersonId"
                                        type="EntityIdType"
                                        minOccurs="0" />
1220                                <xs:element name="Supplier"
                                        type="EntityReferenceType"
                                        minOccurs="0" />
                                </xs:sequence>
                                <xs:attribute name="mandatory"
1225                                type="xs:boolean" />
                                <xs:attribute name="currentlyAssigned"
                                        type="xs:boolean" />
                            </xs:complexType>
                        </xs:element>
1230                    <xs:element name="PositionSpecificCondition"
                            minOccurs="0" maxOccurs="unbounded">
```

```xml
                    <xs:complexType>
                        <xs:sequence>
                            <xs:element name="ConditionCode"
                                type="xs:string"
                                minOccurs="0" />
                            <xs:element
                                name="ConditionDescription"
                                type="xs:string"
                                minOccurs="0" />
                            <xs:element name="ConditionValue"
                                type="xs:string"
                                minOccurs="0" />
                        </xs:sequence>
                    </xs:complexType>
                </xs:element>
            </xs:sequence>
            <xs:attribute name="validFrom" type="xs:date" />
            <xs:attribute name="validTo" type="xs:date" />
        </xs:complexType>
    </xs:element>
    </xs:sequence>
</xs:complexType>
<xs:complexType name="SemCrypt_SalaryType">
    <xs:sequence>
        <xs:element name="PayrollInstructions">
            <xs:complexType>
                <xs:sequence>
                    <xs:element name="PayrollEmployer">
                        <xs:complexType>
                            <xs:sequence>
                                <xs:element name="EmployerId"
                                    type="EntityIdType"
                                    maxOccurs="unbounded" />
                                <xs:element name="EmployerName"
                                    type="xs:string"
                                    minOccurs="0" />
                                <xs:element
                                    name="EmployerGovernmentId"
                                    type="EntityIdType"
                                    minOccurs="0"
                                    maxOccurs="unbounded" />
                            </xs:sequence>
                        </xs:complexType>
                    </xs:element>
                    <xs:element name="PersonInstruction"
                        maxOccurs="unbounded">
                        <xs:complexType>
                            <xs:sequence>
                                <!-- Employee -->
                                <xs:element
                                    name="PayrollPerson">
                                    <xs:complexType>
                                        <xs:sequence>
                                            <xs:element
                                                name="PersonId"
                                                type="EntityIdType"
                                                maxOccurs="unbounded" />
                                            <xs:element
                                                name="PersonName"
                                                type="PersonNameType"
                                                minOccurs="0"
                                                maxOccurs="unbounded" />
                                            <xs:element
                                                name="PersonGovernmentId"
                                                type="EntityIdType"
                                                minOccurs="0"
                                                maxOccurs="unbounded" />
                                        </xs:sequence>
                                    </xs:complexType>
                                </xs:element>
                                <!-- the instruction type -->
                                <xs:element name="Instruction">
                                    <xs:complexType>
                                        <xs:sequence>
                                            <xs:element
                                                name="PaymentDate"
                                                type="xs:date"
                                                minOccurs="0" />
                                            <xs:element
                                                name="Deduction"
                                                type="DeductionType" />
                                            <xs:element
                                                name="Benefit"
```

```
1315                                            type="BenefitType" />
                                       <xs:element
                                           name="Amount"
                                           type="PayrollAmountType"
1320                                        minOccurs="0" />
                                       <xs:element
                                           name="UserArea"
                                           type="UserAreaType"
                                           minOccurs="0" />
                                   </xs:sequence>
1325                               <xs:attribute name="mode"
                                           type="xs:string"
                                           use="required" />
                               </xs:complexType>
                           </xs:element>
1330                       <xs:element name="UserArea"
                                   type="UserAreaType"
                                   minOccurs="0" />
                       </xs:sequence>
                   </xs:complexType>
1335               </xs:element>
                   <!-- the request control totals -->
                   <xs:element name="RequestTotal" minOccurs="0"
                           maxOccurs="unbounded">
                       <xs:complexType>
1340                       <xs:sequence>
                               <xs:element name="Deduction"
                                   type="DeductionType"
                                   minOccurs="0" />
                               <xs:element name="Amount"
1345                               type="PayrollAmountType" />
                           </xs:sequence>
                       </xs:complexType>
                   </xs:element>
                   <xs:element name="UserArea"
1350                       type="UserAreaType" minOccurs="0" />
               </xs:sequence>
               <xs:attribute name="version" type="xs:string" />
               <xs:attribute name="currency" type="xs:string" />
               <!--<xs:attribute ref="xml:lang"/>-->
1355           </xs:complexType>
           </xs:element>
       </xs:sequence>
   </xs:complexType>
   <xs:complexType name="SemCrypt_AssessmentResultsType">
1360       <xs:sequence>
           <xs:element name="AssessmentResult"
               type="AssessmentResultType" maxOccurs="unbounded" />
       </xs:sequence>
   </xs:complexType>
1365 </xs:schema>
```

Appendix B

Service Interfaces

B.1 AuthorizationService.wsdl

In SemCrypt applications, access control is performed on XPath fragments. The authorization service, however, as specified by the SemCrypt Authentication and Authorization Framework (SCAAF) [142], provides a general-purpose authorization operation, which abstracts from actual service implementation. This and the following services are subject to an in-depth discussion in Chapter 6.

Listing B.1: AuthorizationService.wsdl

```
<?xml version="1.0" encoding="UTF-8"?>
<wsdl:definitions xmlns:soap="http://schemas.xmlsoap.org/wsdl/soap/"
    xmlns:tns="http://semcrypt.ec3.at/services/authorize"
    xmlns:wsdl="http://schemas.xmlsoap.org/wsdl/"
    xmlns:xsd="http://www.w3.org/2001/XMLSchema"
    name="AuthorizationService"
    targetNamespace="http://semcrypt.ec3.at/services/authorize">
    <wsdl:types>
        <xsd:schema
            targetNamespace="http://semcrypt.ec3.at/services/authorize">
            <xsd:complexType name="AuthorizationException"></xsd:complexType>
            <xsd:complexType name="CommonResponse"></xsd:complexType>
        </xsd:schema>
    </wsdl:types>
    <wsdl:message name="authorizeRequest">
        <wsdl:part name="token" type="xsd:string" />
        <wsdl:part name="pathValue" type="xsd:string" />
        <wsdl:part name="operation" type="xsd:string"></wsdl:part>
    </wsdl:message>
    <wsdl:message name="authorize_faultMsg">
        <wsdl:part name="fault" type="tns:AuthorizationException" />
    </wsdl:message>
    <wsdl:message name="authorizeResponse">
        <wsdl:part name="response" type="tns:CommonResponse"></wsdl:part>
    </wsdl:message>

    <wsdl:portType name="AuthorizationService">
        <wsdl:operation name="authorize">
            <wsdl:input message="tns:authorizeRequest" />
            <wsdl:output message="tns:authorizeResponse"></wsdl:output>
            <wsdl:fault name="fault"
                message="tns:authorize_faultMsg">
            </wsdl:fault>
        </wsdl:operation>
    </wsdl:portType>

    <wsdl:binding name="AuthorizationServiceSOAP"
```

157

```
              type="tns:AuthorizationService">
              <soap:binding style="document"
40                transport="http://schemas.xmlsoap.org/soap/http" />
              <wsdl:operation name="authorize">
                 <soap:operation
                    soapAction="http://semcrypt.ec3.at/services/authorize/authorize" />
                 <wsdl:input><soap:body use="literal" parts="token pathValue operation" /></wsdl:input>
45               <wsdl:output><soap:body use="literal" parts="response" /></wsdl:output>
              </wsdl:operation>
          </wsdl:binding>
          <wsdl:service name="AuthorizationService">
              <wsdl:port binding="tns:AuthorizationServiceSOAP"
50                name="AuthorizationServiceSOAP">
                 <soap:address
                    location="http://semcrypt.ec3.at/services/authorize" />
              </wsdl:port>
          </wsdl:service>
55      </wsdl:definitions>
```

B.2   CipherService.wsdl

For accessing the encryption/decryption implementation, the cipher service
defined by the SemCrypt Encryption and Decryption Framework (SCEDF)
[144], offers appropriate operations to be included in existing SemCrypt application workflows, as well as general-purpose operations for any client party
that needs to encrypt or decrypt sensitive data.

Listing B.2: CipherService.wsdl

```
<?xml version="1.0" encoding="UTF-8"?>
<wsdl:definitions xmlns:soap="http://schemas.xmlsoap.org/wsdl/soap/"
    xmlns:tns="http://semcrypt.ec3.at/services/cipher"
    xmlns:wsdl="http://schemas.xmlsoap.org/wsdl/"
5   xmlns:xsd="http://www.w3.org/2001/XMLSchema" name="CipherService"
    targetNamespace="http://semcrypt.ec3.at/services/cipher">
    <wsdl:types>
        <xsd:schema
            targetNamespace="http://semcrypt.ec3.at/services/cipher">
10          <xsd:complexType name="CipherException"></xsd:complexType>
        </xsd:schema>
    </wsdl:types>
    <wsdl:message name="encryptRequest">
        <wsdl:part name="token" type="xsd:string" />
15      <wsdl:part name="queryString" type="xsd:string"></wsdl:part>
        <wsdl:part name="plainText" type="xsd:base64Binary"></wsdl:part>
    </wsdl:message>
    <wsdl:message name="decryptRequest">
        <wsdl:part name="token" type="xsd:string"></wsdl:part>
20      <wsdl:part name="queryString" type="xsd:string"></wsdl:part>
        <wsdl:part name="cipherText" type="xsd:base64Binary"></wsdl:part>
    </wsdl:message>
    <wsdl:message name="encryptForAllRequest">
        <wsdl:part name="plainText" type="xsd:base64Binary"></wsdl:part>
25      <wsdl:part name="cipherKey" type="xsd:base64Binary"></wsdl:part>
        <wsdl:part name="algorithm" type="xsd:string"></wsdl:part>
    </wsdl:message>
    <wsdl:message name="decryptForAllRequest">
        <wsdl:part name="cipherText" type="xsd:base64Binary"></wsdl:part>
30      <wsdl:part name="cipherKey" type="xsd:base64Binary"></wsdl:part>
        <wsdl:part name="algorithm" type="xsd:string"></wsdl:part>
    </wsdl:message>
    <wsdl:message name="encryptResponse">
        <wsdl:part name="cipherText" type="xsd:base64Binary"></wsdl:part>
35  </wsdl:message>
    <wsdl:message name="decryptResponse">
        <wsdl:part name="plainText" type="xsd:base64Binary"></wsdl:part>
    </wsdl:message>
    <wsdl:message name="CipherService_faultMsg">
40      <wsdl:part name="fault" type="tns:CipherException"></wsdl:part>
    </wsdl:message>

    <wsdl:portType name="CipherService">
```

```
        <wsdl:operation name="encrypt">
            <wsdl:input message="tns:encryptRequest" />
            <wsdl:output message="tns:encryptResponse" />
            <!-- <wsdl:fault name="fault" message="tns:encrypt_faultMsg"></wsdl:fault> -->
            <wsdl:fault name="fault" message="tns:CipherService_faultMsg"></wsdl:fault>
        </wsdl:operation>
        <wsdl:operation name="decrypt">
            <wsdl:input message="tns:decryptRequest"></wsdl:input>
            <wsdl:output message="tns:decryptResponse"></wsdl:output>
            <!-- <wsdl:fault name="fault" message="tns:decrypt_faultMsg"></wsdl:fault> -->
            <wsdl:fault name="fault" message="tns:CipherService_faultMsg"></wsdl:fault>
        </wsdl:operation>
        <wsdl:operation name="encryptForAll">
            <wsdl:input message="tns:encryptForAllRequest"></wsdl:input>
            <wsdl:output message="tns:encryptResponse"></wsdl:output>
            <!-- <wsdl:fault name="fault" message="tns:encrypt_faultMsg"></wsdl:fault> -->
            <wsdl:fault name="fault"
                message="tns:CipherService_faultMsg">
            </wsdl:fault>
        </wsdl:operation>
        <wsdl:operation name="decryptForAll">
            <wsdl:input message="tns:decryptForAllRequest"></wsdl:input>
            <wsdl:output message="tns:decryptResponse"></wsdl:output>
            <!-- <wsdl:fault name="fault" message="tns:decrypt_faultMsg"></wsdl:fault> -->
            <wsdl:fault name="fault"
                message="tns:CipherService_faultMsg">
            </wsdl:fault>
        </wsdl:operation>
    </wsdl:portType>

    <wsdl:binding name="CipherServiceSOAP" type="tns:CipherService">
        <soap:binding style="document"
            transport="http://schemas.xmlsoap.org/soap/http" />
        <wsdl:operation name="encrypt">
            <soap:operation
                soapAction="http://semcrypt.ec3.at/services/cipher/encrypt" />
            <wsdl:input><soap:body use="literal" /></wsdl:input>
            <wsdl:output><soap:body use="literal" /></wsdl:output>
        </wsdl:operation>
        <wsdl:operation name="decrypt">
            <soap:operation
                soapAction="http://semcrypt.ec3.at/services/cipher/decrypt" />
            <wsdl:input><soap:body use="literal" /></wsdl:input>
            <wsdl:output><soap:body use="literal" /></wsdl:output>
        </wsdl:operation>
        <wsdl:operation name="encryptForAll">
            <soap:operation
                soapAction="http://semcrypt.ec3.at/services/cipher/encryptForAll" />
            <wsdl:input><soap:body use="literal" /></wsdl:input>
            <wsdl:output><soap:body use="literal" /></wsdl:output>
        </wsdl:operation>
        <wsdl:operation name="decryptForAll">
            <soap:operation
                soapAction="http://semcrypt.ec3.at/services/cipher/decryptForAll" />
            <wsdl:input><soap:body use="literal" /></wsdl:input>
            <wsdl:output><soap:body use="literal" /></wsdl:output>
        </wsdl:operation>
    </wsdl:binding>

    <wsdl:service name="CipherService">
        <wsdl:port binding="tns:CipherServiceSOAP"
            name="CipherServiceSOAP">
            <soap:address
                location="http://semcrypt.ec3.at/services/cipher" />
        </wsdl:port>
    </wsdl:service>
</wsdl:definitions>
```

B.3   StorageService.wsdl

The storage service defined by the SemCrypt Database Framework (SCDBF) [143] is not part of the security server, but intended as gateway to interacting with the unsecured data storage provider. Service implementations should only take care of request/response mapping to the underlying database man-

agement system. Security issues, on the other hand, should be handled by
the secure domain and are thus not subject of the storage service. Declared
operations should be used for direct data access and need not invoke security
services themselves.

Listing B.3: StorageService.wsdl

```
<?xml version="1.0" encoding="UTF-8"?>
<wsdl:definitions xmlns:soap="http://schemas.xmlsoap.org/wsdl/soap/"
    xmlns:tns="http://semcrypt.ec3.at/services/storage"
    xmlns:wsdl="http://schemas.xmlsoap.org/wsdl/"
5   xmlns:xsd="http://www.w3.org/2001/XMLSchema" name="StorageService"
    targetNamespace="http://semcrypt.ec3.at/services/storage">
    <wsdl:types>
        <xsd:schema
            targetNamespace="http://semcrypt.ec3.at/services/storage">
10          <xsd:complexType name="StorageException"></xsd:complexType>
            <xsd:complexType name="CommonResponse"></xsd:complexType>
        </xsd:schema>
    </wsdl:types>
    <wsdl:message name="selectDocumentRequest">
15      <wsdl:part name="token" type="xsd:string" />
        <wsdl:part name="documentId" type="xsd:string"></wsdl:part>
    </wsdl:message>
    <wsdl:message name="selectDocumentResponse">
        <wsdl:part name="documentContent" type="xsd:string"></wsdl:part>
20  </wsdl:message>
    <wsdl:message name="updateDocumentRequest">
        <wsdl:part name="token" type="xsd:string"></wsdl:part>
        <wsdl:part name="documentId" type="xsd:string"></wsdl:part>
        <wsdl:part name="documentContent" type="xsd:string"></wsdl:part>
25  </wsdl:message>
    <wsdl:message name="insertDocumentRequest">
        <wsdl:part name="token" type="xsd:string"></wsdl:part>
        <wsdl:part name="documentId" type="xsd:string"></wsdl:part>
        <wsdl:part name="documentContent" type="xsd:string"></wsdl:part>
30  </wsdl:message>
    <wsdl:message name="deleteDocumentRequest">
        <wsdl:part name="token" type="xsd:string"></wsdl:part>
        <wsdl:part name="documentId" type="xsd:string"></wsdl:part>
    </wsdl:message>
35  <wsdl:message name="queryRequest">
        <wsdl:part name="token" type="xsd:string"></wsdl:part>
        <wsdl:part name="documentId" type="xsd:string"></wsdl:part>
        <wsdl:part name="queryRequest" type="xsd:string"></wsdl:part>
    </wsdl:message>
40  <wsdl:message name="queryResponse">
        <wsdl:part name="queryResponse" type="xsd:string"></wsdl:part>
    </wsdl:message>
    <wsdl:message name="updateRequest">
        <wsdl:part name="token" type="xsd:string"></wsdl:part>
45      <wsdl:part name="documentId" type="xsd:string"></wsdl:part>
        <wsdl:part name="updateRequest" type="xsd:string"></wsdl:part>
    </wsdl:message>
    <wsdl:message name="StorageService_commonResponse">
        <wsdl:part name="response" type="tns:CommonResponse" />
50  </wsdl:message>
    <wsdl:message name="StorageService_faultMsg">
        <wsdl:part name="fault" type="tns:StorageException"></wsdl:part>
    </wsdl:message>
    <wsdl:portType name="StorageService">
55      <wsdl:operation name="selectDocument">
            <wsdl:input message="tns:selectDocumentRequest" />
            <wsdl:output message="tns:selectDocumentResponse"></wsdl:output>
            <!-- <wsdl:fault name="fault" message="tns:selectDocument_faultMsg"></wsdl:fault> -->
            <wsdl:fault name="fault"
60              message="tns:StorageService_faultMsg">
            </wsdl:fault>
        </wsdl:operation>
        <wsdl:operation name="insertDocument">
            <wsdl:input message="tns:insertDocumentRequest"></wsdl:input>
65          <!-- <wsdl:fault name="fault" message="tns:insertDocument_faultMsg"></wsdl:fault> -->
            <wsdl:output
                message="tns:StorageService_commonResponse">
            </wsdl:output>
            <wsdl:fault name="fault"
70              message="tns:StorageService_faultMsg">
            </wsdl:fault>
        </wsdl:operation>
        <wsdl:operation name="updateDocument">
```

```
            <wsdl:input message="tns:updateDocumentRequest"></wsdl:input>
            <!-- <wsdl:fault name="fault" message="tns:updateDocument_faultMsg"></wsdl:fault> -->
            <wsdl:output
                message="tns:StorageService_commonResponse">
            </wsdl:output>
            <wsdl:fault name="fault"
                message="tns:StorageService_faultMsg">
            </wsdl:fault>
        </wsdl:operation>
        <wsdl:operation name="deleteDocument">
            <wsdl:input message="tns:deleteDocumentRequest"></wsdl:input>
            <!-- <wsdl:fault name="fault" message="tns:deleteDocument_faultMsg"></wsdl:fault> -->
            <wsdl:output
                message="tns:StorageService_commonResponse">
            </wsdl:output>
            <wsdl:fault name="fault"
                message="tns:StorageService_faultMsg">
            </wsdl:fault>
        </wsdl:operation>
        <wsdl:operation name="query">
            <wsdl:input message="tns:queryRequest"></wsdl:input>
            <wsdl:output message="tns:queryResponse"></wsdl:output>
            <wsdl:fault name="fault"
                message="tns:StorageService_faultMsg">
            </wsdl:fault>
        </wsdl:operation>
        <wsdl:operation name="update">
            <wsdl:input message="tns:updateRequest"></wsdl:input>
            <wsdl:output
                message="tns:StorageService_commonResponse">
            </wsdl:output>
            <wsdl:fault name="fault"
                message="tns:StorageService_faultMsg">
            </wsdl:fault>
        </wsdl:operation>
    </wsdl:portType>
    <wsdl:binding name="StorageServiceSOAP"
        type="tns:StorageService">
        <soap:binding style="document"
            transport="http://schemas.xmlsoap.org/soap/http" />
        <wsdl:operation name="selectDocument">
            <soap:operation
                soapAction="http://semcrypt.ec3.at/services/storage/selectDocument" />
            <wsdl:input>
                <soap:body use="literal" />
            </wsdl:input>
            <wsdl:output>
                <soap:body use="literal" />
            </wsdl:output>
        </wsdl:operation>
        <wsdl:operation name="insertDocument">
            <soap:operation
                soapAction="http://semcrypt.ec3.at/services/storage/insertDocument" />
            <wsdl:input>
                <soap:body use="literal" />
            </wsdl:input>
            <wsdl:output>
                <soap:body use="literal" />
            </wsdl:output>
        </wsdl:operation>
        <wsdl:operation name="updateDocument">
            <soap:operation
                soapAction="http://semcrypt.ec3.at/services/storage/updateDocument" />
            <wsdl:input>
                <soap:body use="literal" />
            </wsdl:input>
            <wsdl:output>
                <soap:body use="literal" />
            </wsdl:output>
        </wsdl:operation>
        <wsdl:operation name="deleteDocument">
            <soap:operation
                soapAction="http://semcrypt.ec3.at/services/storage/deleteDocument" />
            <wsdl:input>
                <soap:body use="literal" />
            </wsdl:input>
            <wsdl:output>
                <soap:body use="literal" />
            </wsdl:output>
        </wsdl:operation>
        <wsdl:operation name="query">
            <soap:operation
                soapAction="http://semcrypt.ec3.at/services/storage/query" />
```

```
                   <wsdl:input>
                       <soap:body use="literal" />
                   </wsdl:input>
160                <wsdl:output>
                       <soap:body use="literal" />
                   </wsdl:output>
               </wsdl:operation>
               <wsdl:operation name="update">
165                <soap:operation
                       soapAction="http://semcrypt.ec3.at/services/storage/update" />
                   <wsdl:input>
                       <soap:body use="literal" />
                   </wsdl:input>
170                <wsdl:output>
                       <soap:body use="literal" />
                   </wsdl:output>
               </wsdl:operation>
           </wsdl:binding>
175        <wsdl:service name="StorageService">
               <wsdl:port binding="tns:StorageServiceSOAP"
                   name="StorageServiceSOAP">
                   <soap:address
                       location="http://semcrypt.ec3.at/services/storage" />
180            </wsdl:port>
           </wsdl:service>
       </wsdl:definitions>
```

B.4   SemCryptService.wsdl

For application developers, who do not want to take care of service interaction configuration, the SemCrypt service contained by the SemCrypt Web Services Framework (SCWSF) [148] and its reference implementation respectively, offers all operations required in typical SemCrypt application settings. Access to fine-granular services may be abstracted through SemCrypt service deployment and configuration.

Listing B.4: SemCryptService.wsdl

```
   <?xml version="1.0" encoding="UTF-8"?>
   <wsdl:definitions xmlns:soap="http://schemas.xmlsoap.org/wsdl/soap/"
       xmlns:tns="http://semcrypt.ec3.at/services/types"
       xmlns:wsdl="http://schemas.xmlsoap.org/wsdl/"
5      xmlns:xsd="http://www.w3.org/2001/XMLSchema"
       name="SemCryptService"
       targetNamespace="http://semcrypt.ec3.at/services/types">
       <wsdl:types>
           <xsd:schema
10             targetNamespace="http://semcrypt.ec3.at/services/types">
               <xsd:complexType name="SemCryptException"></xsd:complexType>
               <xsd:complexType name="CommonResponse"></xsd:complexType>
           </xsd:schema>
       </wsdl:types>
15     <wsdl:message name="queryRequest">
           <wsdl:part name="token" type="xsd:string"></wsdl:part>
           <wsdl:part name="documentId" type="xsd:string"></wsdl:part>
           <wsdl:part name="queryString" type="xsd:string"></wsdl:part>
       </wsdl:message>
20     <wsdl:message name="queryResponse">
           <wsdl:part name="result" type="xsd:string"></wsdl:part>
       </wsdl:message>
       <wsdl:message name="updateRequest">
           <wsdl:part name="token" type="xsd:string"></wsdl:part>
25         <wsdl:part name="documentId" type="xsd:string"></wsdl:part>
           <wsdl:part name="updateString" type="xsd:string"></wsdl:part>
       </wsdl:message>
       <wsdl:message name="SemCryptService_insertDocumentRequest">
           <wsdl:part name="token" type="xsd:string"></wsdl:part>
30         <wsdl:part name="documentId" type="xsd:string"></wsdl:part>
           <wsdl:part name="content" type="xsd:string"></wsdl:part>
           <wsdl:part name="schemaId" type="xsd:string"></wsdl:part>
       </wsdl:message>
```

```
<wsdl:message name="SemCryptService_updateDocumentRequest">
    <wsdl:part name="token" type="xsd:string"></wsdl:part>
    <wsdl:part name="documentId" type="xsd:string"></wsdl:part>
    <wsdl:part name="content" type="xsd:string"></wsdl:part>
</wsdl:message>
<wsdl:message name="SemCryptService_deleteDocumentRequest">
    <wsdl:part name="token" type="xsd:string"></wsdl:part>
    <wsdl:part name="documentId" type="xsd:string"></wsdl:part>
</wsdl:message>
<wsdl:message name="SemCryptService_selectDocumentRequest">
    <wsdl:part name="token" type="xsd:string"></wsdl:part>
    <wsdl:part name="documentId" type="xsd:string"></wsdl:part>
</wsdl:message>
<wsdl:message name="SemCryptService_selectDocumentResponse">
    <wsdl:part name="content" type="xsd:string"></wsdl:part>
</wsdl:message>
<wsdl:message name="SemCryptService_insertSchemaRequest">
    <wsdl:part name="token" type="xsd:string"></wsdl:part>
    <wsdl:part name="schemaId" type="xsd:string"></wsdl:part>
    <wsdl:part name="content" type="xsd:string"></wsdl:part>
</wsdl:message>
<wsdl:message name="SemCryptService_updateSchemaRequest">
    <wsdl:part name="token" type="xsd:string"></wsdl:part>
    <wsdl:part name="schemaId" type="xsd:string"></wsdl:part>
    <wsdl:part name="content" type="xsd:string"></wsdl:part>
</wsdl:message>
<wsdl:message name="SemCryptService_deleteSchemaRequest">
    <wsdl:part name="token" type="xsd:string"></wsdl:part>
    <wsdl:part name="schemaId" type="xsd:string"></wsdl:part>
</wsdl:message>
<wsdl:message name="SemCryptService_selectSchemaRequest">
    <wsdl:part name="token" type="xsd:string"></wsdl:part>
    <wsdl:part name="schemaId" type="xsd:string"></wsdl:part>
</wsdl:message>
<wsdl:message name="SemCryptService_selectSchemaResponse">
    <wsdl:part name="content" type="xsd:string"></wsdl:part>
</wsdl:message>
<wsdl:message name="SemCryptServiceCommonResponse">
    <wsdl:part name="response" type="tns:CommonResponse"></wsdl:part>
</wsdl:message>
<wsdl:message name="SemCryptService_faultMsg">
    <wsdl:part name="fault" type="tns:SemCryptException"></wsdl:part>
</wsdl:message>
<wsdl:portType name="SemCryptService">
    <wsdl:operation name="selectDocument">
        <wsdl:input
            message="tns:SemCryptService_selectDocumentRequest">
        </wsdl:input>
        <wsdl:output
            message="tns:SemCryptService_selectDocumentResponse">
        </wsdl:output>
        <wsdl:fault name="fault"
            message="tns:SemCryptService_faultMsg">
        </wsdl:fault>
    </wsdl:operation>
    <wsdl:operation name="deleteDocument">
        <wsdl:input
            message="tns:SemCryptService_deleteDocumentRequest">
        </wsdl:input>
        <wsdl:output
            message="tns:SemCryptServiceCommonResponse">
        </wsdl:output>
        <wsdl:fault name="fault"
            message="tns:SemCryptService_faultMsg">
        </wsdl:fault>
    </wsdl:operation>
    <wsdl:operation name="insertDocument">
        <wsdl:input
            message="tns:SemCryptService_insertDocumentRequest">
        </wsdl:input>
        <wsdl:output
            message="tns:SemCryptServiceCommonResponse">
        </wsdl:output>
        <wsdl:fault name="fault"
            message="tns:SemCryptService_faultMsg">
        </wsdl:fault>
    </wsdl:operation>
    <wsdl:operation name="insertDocumentForUser">
        <wsdl:input
            message="tns:SemCryptService_insertDocumentRequest">
        </wsdl:input>
        <wsdl:output
            message="tns:SemCryptServiceCommonResponse">
```

```
                    </wsdl:output>
                    <wsdl:fault name="fault"
                        message="tns:SemCryptService_faultMsg">
120                 </wsdl:fault>
                </wsdl:operation>
                <wsdl:operation name="insertDocumentForRole">
                    <wsdl:input
                        message="tns:SemCryptService_insertDocumentRequest">
125                 </wsdl:input>
                    <wsdl:output
                        message="tns:SemCryptServiceCommonResponse">
                    </wsdl:output>
                    <wsdl:fault name="fault"
130                     message="tns:SemCryptService_faultMsg">
                    </wsdl:fault>
                </wsdl:operation>
                <wsdl:operation name="updateDocument">
                    <wsdl:input
135                     message="tns:SemCryptService_updateDocumentRequest">
                    </wsdl:input>
                    <wsdl:output
                        message="tns:SemCryptServiceCommonResponse">
                    </wsdl:output>
140                 <wsdl:fault name="fault"
                        message="tns:SemCryptService_faultMsg">
                    </wsdl:fault>
                </wsdl:operation>
                <wsdl:operation name="insertSchema">
145                 <wsdl:input
                        message="tns:SemCryptService_insertSchemaRequest">
                    </wsdl:input>
                    <wsdl:output
                        message="tns:SemCryptServiceCommonResponse">
150                 </wsdl:output>
                    <wsdl:fault name="fault"
                        message="tns:SemCryptService_faultMsg">
                    </wsdl:fault>
                </wsdl:operation>
155             <wsdl:operation name="insertSchemaForUser">
                    <wsdl:input
                        message="tns:SemCryptService_insertSchemaRequest">
                    </wsdl:input>
                    <wsdl:output
160                     message="tns:SemCryptServiceCommonResponse">
                    </wsdl:output>
                    <wsdl:fault name="fault"
                        message="tns:SemCryptService_faultMsg">
                    </wsdl:fault>
165             </wsdl:operation>
                <wsdl:operation name="insertSchemaForRole">
                    <wsdl:input
                        message="tns:SemCryptService_insertSchemaRequest">
                    </wsdl:input>
170                 <wsdl:output
                        message="tns:SemCryptServiceCommonResponse">
                    </wsdl:output>
                    <wsdl:fault name="fault"
                        message="tns:SemCryptService_faultMsg">
175                 </wsdl:fault>
                </wsdl:operation>
                <wsdl:operation name="updateSchema">
                    <wsdl:input
                        message="tns:SemCryptService_updateSchemaRequest">
180                 </wsdl:input>
                    <wsdl:output
                        message="tns:SemCryptServiceCommonResponse">
                    </wsdl:output>
                    <wsdl:fault name="fault"
185                     message="tns:SemCryptService_faultMsg">
                    </wsdl:fault>
                </wsdl:operation>
                <wsdl:operation name="deleteSchema">
                    <wsdl:input
190                     message="tns:SemCryptService_deleteSchemaRequest">
                    </wsdl:input>
                    <wsdl:output
                        message="tns:SemCryptServiceCommonResponse">
                    </wsdl:output>
195                 <wsdl:fault name="fault"
                        message="tns:SemCryptService_faultMsg">
                    </wsdl:fault>
                </wsdl:operation>
                <wsdl:operation name="selectSchema">
```

```
        <wsdl:input
            message="tns:SemCryptService_selectSchemaRequest">
        </wsdl:input>
        <wsdl:output
            message="tns:SemCryptService_selectSchemaResponse">
        </wsdl:output>
        <wsdl:fault name="fault"
            message="tns:SemCryptService_faultMsg">
        </wsdl:fault>
    </wsdl:operation>
    <wsdl:operation name="query">
        <wsdl:input message="tns:queryRequest"></wsdl:input>
        <wsdl:output message="tns:queryResponse"></wsdl:output>
        <wsdl:fault name="fault"
            message="tns:SemCryptService_faultMsg">
        </wsdl:fault>
    </wsdl:operation>
    <wsdl:operation name="update">
        <wsdl:input message="tns:updateRequest"></wsdl:input>
        <wsdl:output
            message="tns:SemCryptServiceCommonResponse">
        </wsdl:output>
        <wsdl:fault name="fault"
            message="tns:SemCryptService_faultMsg">
        </wsdl:fault>
    </wsdl:operation>
</wsdl:portType>
<wsdl:binding name="SemCryptServiceSOAP"
    type="tns:SemCryptService">
    <soap:binding style="document"
        transport="http://schemas.xmlsoap.org/soap/http" />
    <wsdl:operation name="selectDocument">
        <soap:operation
            soapAction="http://semcrypt.ec3.at/services/types/selectDocument" />
        <wsdl:input>
            <soap:body use="literal" />
        </wsdl:input>
        <wsdl:output>
            <soap:body use="literal" />
        </wsdl:output>
    </wsdl:operation>
    <wsdl:operation name="deleteDocument">
        <soap:operation
            soapAction="http://semcrypt.ec3.at/services/types/deleteDocument" />
        <wsdl:input>
            <soap:body use="literal" />
        </wsdl:input>
        <wsdl:output>
            <soap:body use="literal" />
        </wsdl:output>
    </wsdl:operation>
    <wsdl:operation name="insertDocument">
        <soap:operation
            soapAction="http://semcrypt.ec3.at/services/types/insertDocument" />
        <wsdl:input>
            <soap:body use="literal" />
        </wsdl:input>
        <wsdl:output>
            <soap:body use="literal" />
        </wsdl:output>
    </wsdl:operation>
    <wsdl:operation name="insertDocumentForRole">
        <soap:operation
            soapAction="http://semcrypt.ec3.at/services/types/insertDocumentForRole" />
        <wsdl:input>
            <soap:body use="literal" />
        </wsdl:input>
        <wsdl:output>
            <soap:body use="literal" />
        </wsdl:output>
    </wsdl:operation>
    <wsdl:operation name="insertDocumentForUser">
        <soap:operation
            soapAction="http://semcrypt.ec3.at/services/types/insertDocumentForUser" />
        <wsdl:input>
            <soap:body use="literal" />
        </wsdl:input>
        <wsdl:output>
            <soap:body use="literal" />
        </wsdl:output>
    </wsdl:operation>
    <wsdl:operation name="updateDocument">
        <soap:operation
```

```
                         soapAction="http://semcrypt.ec3.at/services/types/updateDocument" />
                         <wsdl:input>
285                          <soap:body use="literal" />
                         </wsdl:input>
                         <wsdl:output>
                             <soap:body use="literal" />
                         </wsdl:output>
290                      </wsdl:operation>
                         <wsdl:operation name="selectSchema">
                         <soap:operation
                             soapAction="http://semcrypt.ec3.at/services/types/selectSchema" />
                         <wsdl:input>
295                          <soap:body use="literal" />
                         </wsdl:input>
                         <wsdl:output>
                             <soap:body use="literal" />
                         </wsdl:output>
300                      </wsdl:operation>
                         <wsdl:operation name="deleteSchema">
                         <soap:operation
                             soapAction="http://semcrypt.ec3.at/services/types/deleteSchema" />
                         <wsdl:input>
305                          <soap:body use="literal" />
                         </wsdl:input>
                         <wsdl:output>
                             <soap:body use="literal" />
                         </wsdl:output>
310                      </wsdl:operation>
                         <wsdl:operation name="insertSchema">
                         <soap:operation
                             soapAction="http://semcrypt.ec3.at/services/types/insertSchema" />
                         <wsdl:input>
315                          <soap:body use="literal" />
                         </wsdl:input>
                         <wsdl:output>
                             <soap:body use="literal" />
                         </wsdl:output>
320                      </wsdl:operation>
                         <wsdl:operation name="insertSchemaForRole">
                         <soap:operation
                             soapAction="http://semcrypt.ec3.at/services/types/insertSchemaForRole" />
                         <wsdl:input>
325                          <soap:body use="literal" />
                         </wsdl:input>
                         <wsdl:output>
                             <soap:body use="literal" />
                         </wsdl:output>
330                      </wsdl:operation>
                         <wsdl:operation name="insertSchemaForUser">
                         <soap:operation
                             soapAction="http://semcrypt.ec3.at/services/types/insertSchemaForUser" />
                         <wsdl:input>
335                          <soap:body use="literal" />
                         </wsdl:input>
                         <wsdl:output>
                             <soap:body use="literal" />
                         </wsdl:output>
340                      </wsdl:operation>
                         <wsdl:operation name="updateSchema">
                         <soap:operation
                             soapAction="http://semcrypt.ec3.at/services/types/updateSchema" />
                         <wsdl:input>
345                          <soap:body use="literal" />
                         </wsdl:input>
                         <wsdl:output>
                             <soap:body use="literal" />
                         </wsdl:output>
350                      </wsdl:operation>
                         <wsdl:operation name="query">
                         <soap:operation
                             soapAction="http://semcrypt.ec3.at/services/types/query" />
                         <wsdl:input>
355                          <soap:body use="literal" />
                         </wsdl:input>
                         <wsdl:output>
                             <soap:body use="literal" />
                         </wsdl:output>
360                      </wsdl:operation>
                         <wsdl:operation name="update">
                         <soap:operation
                             soapAction="http://semcrypt.ec3.at/services/types/update" />
                         <wsdl:input>
365                          <soap:body use="literal" />
```

```
            </wsdl:input>
            <wsdl:output>
                <soap:body use="literal" />
            </wsdl:output>
        </wsdl:operation>
    </wsdl:binding>
    <wsdl:service name="SemCryptService">
        <wsdl:port binding="tns:SemCryptServiceSOAP"
            name="SemCryptServiceSOAP">
            <soap:address
                location="http://semcrypt.ec3.at/services/types" />
        </wsdl:port>
    </wsdl:service>
</wsdl:definitions>
```

B.5 ManagementService.wsdl

In order to facilitate application deployment and runtime configuration, the
management service as specified by the SemCrypt Web Services Framework
(SCWSF) [148] provides operations for selecting and modifying access control
policies, encryption instructions, as well as application users and user groups,
also referred to as roles.

Listing B.5: ManagementService.wsdl

```
<?xml version="1.0" encoding="UTF-8"?>
<wsdl:definitions xmlns:soap="http://schemas.xmlsoap.org/wsdl/soap/"
    xmlns:tns="http://semcrypt.ec3.at/services/management"
    xmlns:wsdl="http://schemas.xmlsoap.org/wsdl/"
    xmlns:xsd="http://www.w3.org/2001/XMLSchema"
    name="ManagementService"
    targetNamespace="http://semcrypt.ec3.at/services/management">
    <wsdl:types>
        <xsd:schema
            targetNamespace="http://semcrypt.ec3.at/services/management">
            <xsd:complexType name="CommonResponse"></xsd:complexType>
            <xsd:complexType name="ManagementException"></xsd:complexType>
            <xsd:complexType name="User"></xsd:complexType>
            <xsd:complexType name="Role"></xsd:complexType>
        </xsd:schema>
    </wsdl:types>
    <wsdl:message name="insertPolicyRequest">
        <wsdl:part name="token" type="xsd:string"></wsdl:part>
        <wsdl:part name="policyData" type="xsd:base64Binary" />
    </wsdl:message>
    <wsdl:message name="insertSpecificPolicyRequest">
        <wsdl:part name="token" type="xsd:string"></wsdl:part>
        <wsdl:part name="policyData" type="xsd:base64Binary"></wsdl:part>
        <wsdl:part name="documentId" type="xsd:string"></wsdl:part>
    </wsdl:message>
    <wsdl:message name="updatePolicyRequest">
        <wsdl:part name="token" type="xsd:string"></wsdl:part>
        <wsdl:part name="policyData" type="xsd:base64Binary"></wsdl:part>
    </wsdl:message>
    <wsdl:message name="updateSpecificPolicyRequest">
        <wsdl:part name="token" type="xsd:string"></wsdl:part>
        <wsdl:part name="policyData" type="xsd:base64Binary"></wsdl:part>
        <wsdl:part name="documentId" type="xsd:string"></wsdl:part>
    </wsdl:message>
    <wsdl:message name="deletePolicyRequest">
        <wsdl:part name="token" type="xsd:string"></wsdl:part>
    </wsdl:message>
    <wsdl:message name="deleteSpecificPolicyRequest">
        <wsdl:part name="token" type="xsd:string"></wsdl:part>
        <wsdl:part name="documentId" type="xsd:string"></wsdl:part>
    </wsdl:message>
    <wsdl:message name="selectPolicyRequest">
        <wsdl:part name="documentId" type="xsd:string"></wsdl:part>
    </wsdl:message>
    <wsdl:message name="selectPolicyResponse">
        <wsdl:part name="policyData" type="xsd:base64Binary"></wsdl:part>
    </wsdl:message>
```

```
     <wsdl:message name="selectSpecificPolicyRequest">
       <wsdl:part name="token" type="xsd:string"></wsdl:part>
50     <wsdl:part name="documentId" type="xsd:string"></wsdl:part>
     </wsdl:message>
     <wsdl:message name="insertXCipherRequest">
       <wsdl:part name="token" type="xsd:string"></wsdl:part>
       <wsdl:part name="xcipherData" type="xsd:base64Binary"></wsdl:part>
55   </wsdl:message>
     <wsdl:message name="insertSpecificXCipherRequest">
       <wsdl:part name="token" type="xsd:string"></wsdl:part>
       <wsdl:part name="xcipherData" type="xsd:base64Binary"></wsdl:part>
       <wsdl:part name="documentId" type="xsd:string"></wsdl:part>
60   </wsdl:message>
     <wsdl:message name="updateXCipherRequest">
       <wsdl:part name="token" type="xsd:string"></wsdl:part>
       <wsdl:part name="xcipherData" type="xsd:base64Binary"></wsdl:part>
     </wsdl:message>
65   <wsdl:message name="updateSpecificXCipherRequest">
       <wsdl:part name="token" type="xsd:string"></wsdl:part>
       <wsdl:part name="xcipherData" type="xsd:base64Binary"></wsdl:part>
       <wsdl:part name="documentId" type="xsd:string"></wsdl:part>
     </wsdl:message>
70   <wsdl:message name="deleteXCipherRequest">
       <wsdl:part name="token" type="xsd:string"></wsdl:part>
     </wsdl:message>
     <wsdl:message name="deleteSpecificXCipherRequest">
       <wsdl:part name="token" type="xsd:string"></wsdl:part>
75     <wsdl:part name="documentId" type="xsd:string"></wsdl:part>
     </wsdl:message>
     <wsdl:message name="selectXCipherRequest">
       <wsdl:part name="token" type="xsd:string"></wsdl:part>
     </wsdl:message>
80   <wsdl:message name="selectXCipherResponse">
       <wsdl:part name="xcipherData" type="xsd:base64Binary"></wsdl:part>
     </wsdl:message>
     <wsdl:message name="selectSpecificXCipherRequest">
       <wsdl:part name="token" type="xsd:string"></wsdl:part>
85     <wsdl:part name="documentId" type="xsd:string"></wsdl:part>
     </wsdl:message>
     <wsdl:message name="insertKeyRequest">
       <wsdl:part name="token" type="xsd:string"></wsdl:part>
       <wsdl:part name="keyData" type="xsd:base64Binary"></wsdl:part>
90     <wsdl:part name="keyId" type="xsd:string"></wsdl:part>
     </wsdl:message>
     <wsdl:message name="updateKeyRequest">
       <wsdl:part name="token" type="xsd:string"></wsdl:part>
       <wsdl:part name="keyData" type="xsd:base64Binary"></wsdl:part>
95     <wsdl:part name="keyId" type="xsd:string"></wsdl:part>
     </wsdl:message>
     <wsdl:message name="deleteKeyRequest">
       <wsdl:part name="token" type="xsd:string"></wsdl:part>
       <wsdl:part name="keyId" type="xsd:string"></wsdl:part>
100  </wsdl:message>
     <wsdl:message name="selectKeyRequest">
       <wsdl:part name="token" type="xsd:string"></wsdl:part>
       <wsdl:part name="keyId" type="xsd:string"></wsdl:part>
     </wsdl:message>
105  <wsdl:message name="selectKeyResponse">
       <wsdl:part name="keyData" type="xsd:base64Binary"></wsdl:part>
     </wsdl:message>
     <wsdl:message name="insertUserRequest">
       <wsdl:part name="token" type="xsd:string"></wsdl:part>
110    <wsdl:part name="user" type="tns:User"></wsdl:part>
     </wsdl:message>
     <wsdl:message name="updateUserRequest">
       <wsdl:part name="token" type="xsd:string"></wsdl:part>
       <wsdl:part name="user" type="tns:User"></wsdl:part>
115  </wsdl:message>
     <wsdl:message name="selectUserRequest">
       <wsdl:part name="token" type="xsd:string"></wsdl:part>
       <wsdl:part name="user" type="tns:User"></wsdl:part>
     </wsdl:message>
120  <wsdl:message name="selectUserResponse">
       <wsdl:part name="user" type="tns:User"></wsdl:part>
     </wsdl:message>
     <wsdl:message name="deleteUserRequest">
       <wsdl:part name="token" type="xsd:string"></wsdl:part>
125    <wsdl:part name="user" type="tns:User"></wsdl:part>
     </wsdl:message>
     <wsdl:message name="insertRoleRequest">
       <wsdl:part name="token" type="xsd:string"></wsdl:part>
       <wsdl:part name="role" type="tns:Role"></wsdl:part>
130  </wsdl:message>
```

```
<wsdl:message name="updateRoleRequest">
    <wsdl:part name="token" type="xsd:string"></wsdl:part>
    <wsdl:part name="role" type="tns:Role"></wsdl:part>
</wsdl:message>
<wsdl:message name="deleteRoleRequest">
    <wsdl:part name="token" type="xsd:string"></wsdl:part>
    <wsdl:part name="role" type="tns:Role"></wsdl:part>
</wsdl:message>
<wsdl:message name="selectRoleRequest">
    <wsdl:part name="token" type="xsd:string"></wsdl:part>
    <wsdl:part name="role" type="tns:Role"></wsdl:part>
</wsdl:message>
<wsdl:message name="selectRoleResponse">
    <wsdl:part name="role" type="tns:Role"></wsdl:part>
</wsdl:message>
<wsdl:message name="ManagementService_commonResponse">
    <wsdl:part name="response" type="tns:CommonResponse"></wsdl:part>
</wsdl:message>
<wsdl:message name="ManagementService_faultMsg">
    <wsdl:part name="fault" type="tns:ManagementException"></wsdl:part>
</wsdl:message>
<wsdl:portType name="ManagementService">
    <wsdl:operation name="insertPolicy">
        <wsdl:input message="tns:insertPolicyRequest" />
        <wsdl:output
            message="tns:ManagementService_commonResponse" />
        <wsdl:fault name="fault"
            message="tns:ManagementService_faultMsg">
        </wsdl:fault>
    </wsdl:operation>
    <wsdl:operation name="insertDocumentPolicy">
        <wsdl:input message="tns:insertSpecificPolicyRequest"></wsdl:input>
        <wsdl:output
            message="tns:ManagementService_commonResponse">
        </wsdl:output>
        <wsdl:fault name="fault"
            message="tns:ManagementService_faultMsg">
        </wsdl:fault>
    </wsdl:operation>
    <wsdl:operation name="updatePolicy">
        <wsdl:input message="tns:updatePolicyRequest"></wsdl:input>
        <wsdl:output
            message="tns:ManagementService_commonResponse">
        </wsdl:output>
        <wsdl:fault name="fault"
            message="tns:ManagementService_faultMsg">
        </wsdl:fault>
    </wsdl:operation>
    <wsdl:operation name="updateDocumentPolicy">
        <wsdl:input message="tns:updateSpecificPolicyRequest"></wsdl:input>
        <wsdl:output
            message="tns:ManagementService_commonResponse">
        </wsdl:output>
        <wsdl:fault name="fault"
            message="tns:ManagementService_faultMsg">
        </wsdl:fault>
    </wsdl:operation>
    <wsdl:operation name="deletePolicy">
        <wsdl:input message="tns:deletePolicyRequest"></wsdl:input>
        <wsdl:output
            message="tns:ManagementService_commonResponse">
        </wsdl:output>
        <wsdl:fault name="fault"
            message="tns:ManagementService_faultMsg">
        </wsdl:fault>
    </wsdl:operation>
    <wsdl:operation name="deleteDocumentPolicy">
        <wsdl:input message="tns:deleteSpecificPolicyRequest"></wsdl:input>
        <wsdl:output
            message="tns:ManagementService_commonResponse">
        </wsdl:output>
        <wsdl:fault name="fault"
            message="tns:ManagementService_faultMsg">
        </wsdl:fault>
    </wsdl:operation>
    <wsdl:operation name="selectPolicy">
        <wsdl:input message="tns:selectPolicyRequest"></wsdl:input>
        <wsdl:output message="tns:selectPolicyResponse"></wsdl:output>
        <wsdl:fault name="fault"
            message="tns:ManagementService_faultMsg">
        </wsdl:fault>
    </wsdl:operation>
    <wsdl:operation name="selectDocumentPolicy">
```

```
                    <wsdl:input message="tns:selectSpecificPolicyRequest"></wsdl:input>
215                 <wsdl:output message="tns:selectPolicyResponse"></wsdl:output>
                    <wsdl:fault name="fault"
                        message="tns:ManagementService_faultMsg">
                    </wsdl:fault>
                </wsdl:operation>
220             <wsdl:operation name="insertXCipher">
                    <wsdl:input message="tns:insertXCipherRequest"></wsdl:input>
                    <wsdl:output
                        message="tns:ManagementService_commonResponse">
                    </wsdl:output>
225                 <wsdl:fault name="fault"
                        message="tns:ManagementService_faultMsg">
                    </wsdl:fault>
                </wsdl:operation>
                <wsdl:operation name="insertDocumentXCipher">
230                 <wsdl:input message="tns:insertSpecificXCipherRequest"></wsdl:input>
                    <wsdl:output
                        message="tns:ManagementService_commonResponse">
                    </wsdl:output>
                    <wsdl:fault name="fault"
235                     message="tns:ManagementService_faultMsg">
                    </wsdl:fault>
                </wsdl:operation>
                <wsdl:operation name="updateXCipher">
                    <wsdl:input message="tns:updateXCipherRequest"></wsdl:input>
240                 <wsdl:output
                        message="tns:ManagementService_commonResponse">
                    </wsdl:output>
                    <wsdl:fault name="fault"
                        message="tns:ManagementService_faultMsg">
245                 </wsdl:fault>
                </wsdl:operation>
                <wsdl:operation name="updateDocumentXCipher">
                    <wsdl:input message="tns:updateSpecificXCipherRequest"></wsdl:input>
                    <wsdl:output
250                     message="tns:ManagementService_commonResponse">
                    </wsdl:output>
                    <wsdl:fault name="fault"
                        message="tns:ManagementService_faultMsg">
                    </wsdl:fault>
255             </wsdl:operation>
                <wsdl:operation name="deleteXCipher">
                    <wsdl:input message="tns:deleteXCipherRequest"></wsdl:input>
                    <wsdl:output
                        message="tns:ManagementService_commonResponse">
260                 </wsdl:output>
                    <wsdl:fault name="fault"
                        message="tns:ManagementService_faultMsg">
                    </wsdl:fault>
                </wsdl:operation>
265             <wsdl:operation name="deleteDocumentXCipher">
                    <wsdl:input message="tns:deleteSpecificXCipherRequest"></wsdl:input>
                    <wsdl:output
                        message="tns:ManagementService_commonResponse">
                    </wsdl:output>
270                 <wsdl:fault name="fault"
                        message="tns:ManagementService_faultMsg">
                    </wsdl:fault>
                </wsdl:operation>
                <wsdl:operation name="selectXCipher">
275                 <wsdl:input message="tns:selectXCipherRequest"></wsdl:input>
                    <wsdl:output message="tns:selectXCipherResponse"></wsdl:output>
                    <wsdl:fault name="fault"
                        message="tns:ManagementService_faultMsg">
                    </wsdl:fault>
280             </wsdl:operation>
                <wsdl:operation name="selectDocumentXCipher">
                    <wsdl:input message="tns:selectSpecificXCipherRequest"></wsdl:input>
                    <wsdl:output message="tns:selectXCipherResponse"></wsdl:output>
                    <wsdl:fault name="fault"
285                     message="tns:ManagementService_faultMsg">
                    </wsdl:fault>
                </wsdl:operation>
                <wsdl:operation name="insertKey">
                    <wsdl:input message="tns:insertKeyRequest"></wsdl:input>
290                 <wsdl:output
                        message="tns:ManagementService_commonResponse">
                    </wsdl:output>
                    <wsdl:fault name="fault"
                        message="tns:ManagementService_faultMsg">
295                 </wsdl:fault>
                </wsdl:operation>
```

```
<wsdl:operation name="updateKey">
    <wsdl:input message="tns:updateKeyRequest"></wsdl:input>
    <wsdl:output
        message="tns:ManagementService_commonResponse">
    </wsdl:output>
    <wsdl:fault name="fault"
        message="tns:ManagementService_faultMsg">
    </wsdl:fault>
</wsdl:operation>
<wsdl:operation name="deleteKey">
    <wsdl:input message="tns:deleteKeyRequest"></wsdl:input>
    <wsdl:output
        message="tns:ManagementService_commonResponse">
    </wsdl:output>
    <wsdl:fault name="fault"
        message="tns:ManagementService_faultMsg">
    </wsdl:fault>
</wsdl:operation>
<wsdl:operation name="selectKey">
    <wsdl:input message="tns:selectKeyRequest"></wsdl:input>
    <wsdl:output message="tns:selectKeyResponse"></wsdl:output>
    <wsdl:fault name="fault"
        message="tns:ManagementService_faultMsg">
    </wsdl:fault>
</wsdl:operation>
<wsdl:operation name="insertUser">
    <wsdl:input message="tns:insertUserRequest"></wsdl:input>
    <wsdl:output
        message="tns:ManagementService_commonResponse">
    </wsdl:output>
    <wsdl:fault name="fault"
        message="tns:ManagementService_faultMsg">
    </wsdl:fault>
</wsdl:operation>
<wsdl:operation name="updateUser">
    <wsdl:input message="tns:updateUserRequest"></wsdl:input>
    <wsdl:output
        message="tns:ManagementService_commonResponse">
    </wsdl:output>
    <wsdl:fault name="fault"
        message="tns:ManagementService_faultMsg">
    </wsdl:fault>
</wsdl:operation>
<wsdl:operation name="deleteUser">
    <wsdl:input message="tns:selectUserRequest"></wsdl:input>
    <wsdl:output
        message="tns:ManagementService_commonResponse">
    </wsdl:output>
    <wsdl:fault name="fault"
        message="tns:ManagementService_faultMsg">
    </wsdl:fault>
</wsdl:operation>
<wsdl:operation name="insertRole">
    <wsdl:input message="tns:insertRoleRequest"></wsdl:input>
    <wsdl:output
        message="tns:ManagementService_commonResponse">
    </wsdl:output>
    <wsdl:fault name="fault"
        message="tns:ManagementService_faultMsg">
    </wsdl:fault>
</wsdl:operation>
<wsdl:operation name="updateRole">
    <wsdl:input message="tns:updateRoleRequest"></wsdl:input>
    <wsdl:output
        message="tns:ManagementService_commonResponse">
    </wsdl:output>
    <wsdl:fault name="fault"
        message="tns:ManagementService_faultMsg">
    </wsdl:fault>
</wsdl:operation>
<wsdl:operation name="deleteRole">
    <wsdl:input message="tns:deleteRoleRequest"></wsdl:input>
    <wsdl:output
        message="tns:ManagementService_commonResponse">
    </wsdl:output>
    <wsdl:fault name="fault"
        message="tns:ManagementService_faultMsg">
    </wsdl:fault>
</wsdl:operation>
<wsdl:operation name="selectRole">
    <wsdl:input message="tns:selectRoleRequest"></wsdl:input>
    <wsdl:output message="tns:selectRoleResponse"></wsdl:output>
    <wsdl:fault name="fault"
```

```
380                     message="tns:ManagementService_faultMsg">
                    </wsdl:fault>
                </wsdl:operation>
                <wsdl:operation name="selectUser">
                    <wsdl:input message="tns:selectUserRequest"></wsdl:input>
385                 <wsdl:output message="tns:selectUserResponse"></wsdl:output>
                    <wsdl:fault name="fault"
                        message="tns:ManagementService_faultMsg">
                    </wsdl:fault>
                </wsdl:operation>
390     </wsdl:portType>
        <wsdl:binding name="ManagementServiceSOAP"
            type="tns:ManagementService">
            <soap:binding style="document"
                transport="http://schemas.xmlsoap.org/soap/http" />
395         <wsdl:operation name="insertPolicy">
                <soap:operation
                    soapAction="http://semcrypt.ec3.at/services/management/insertPolicy" />
                <wsdl:input>
                    <soap:body use="literal" />
400             </wsdl:input>
                <wsdl:output>
                    <soap:body use="literal" />
                </wsdl:output>
            </wsdl:operation>
405         <wsdl:operation name="insertDocumentPolicy">
                <soap:operation
                    soapAction="http://semcrypt.ec3.at/services/management/insertPolicyForRole" />
                <wsdl:input>
                    <soap:body use="literal" />
410             </wsdl:input>
                <wsdl:output>
                    <soap:body use="literal" />
                </wsdl:output>
            </wsdl:operation>
415         <wsdl:operation name="updatePolicy">
                <soap:operation
                    soapAction="http://semcrypt.ec3.at/services/management/updatePolicy" />
                <wsdl:input>
                    <soap:body use="literal" />
420             </wsdl:input>
                <wsdl:output>
                    <soap:body use="literal" />
                </wsdl:output>
            </wsdl:operation>
425         <wsdl:operation name="updateDocumentPolicy">
                <soap:operation
                    soapAction="http://semcrypt.ec3.at/services/management/updatePolicyForRole" />
                <wsdl:input>
                    <soap:body use="literal" />
430             </wsdl:input>
                <wsdl:output>
                    <soap:body use="literal" />
                </wsdl:output>
            </wsdl:operation>
435         <wsdl:operation name="deletePolicy">
                <soap:operation
                    soapAction="http://semcrypt.ec3.at/services/management/deletePolicy" />
                <wsdl:input>
                    <soap:body use="literal" />
440             </wsdl:input>
                <wsdl:output>
                    <soap:body use="literal" />
                </wsdl:output>
            </wsdl:operation>
445         <wsdl:operation name="deleteDocumentPolicy">
                <soap:operation
                    soapAction="http://semcrypt.ec3.at/services/management/deletePolicyForUser" />
                <wsdl:input>
                    <soap:body use="literal" />
450             </wsdl:input>
                <wsdl:output>
                    <soap:body use="literal" />
                </wsdl:output>
            </wsdl:operation>
455         <wsdl:operation name="selectPolicy">
                <soap:operation
                    soapAction="http://semcrypt.ec3.at/services/management/selectPolicy" />
                <wsdl:input>
                    <soap:body use="literal" />
460             </wsdl:input>
                <wsdl:output>
                    <soap:body use="literal" />
```

```
            </wsdl:output>
        </wsdl:operation>
        <wsdl:operation name="selectDocumentPolicy">
            <soap:operation
                soapAction="http://semcrypt.ec3.at/services/management/selectPolicyForRole" />
            <wsdl:input>
                <soap:body use="literal" />
            </wsdl:input>
            <wsdl:output>
                <soap:body use="literal" />
            </wsdl:output>
        </wsdl:operation>
        <wsdl:operation name="insertXCipher">
            <soap:operation
                soapAction="http://semcrypt.ec3.at/services/management/insertXCipher" />
            <wsdl:input>
                <soap:body use="literal" />
            </wsdl:input>
            <wsdl:output>
                <soap:body use="literal" />
            </wsdl:output>
        </wsdl:operation>
        <wsdl:operation name="insertDocumentXCipher">
            <soap:operation
                soapAction="http://semcrypt.ec3.at/services/management/insertXCipherForUser" />
            <wsdl:input>
                <soap:body use="literal" />
            </wsdl:input>
            <wsdl:output>
                <soap:body use="literal" />
            </wsdl:output>
        </wsdl:operation>
        <wsdl:operation name="updateXCipher">
            <soap:operation
                soapAction="http://semcrypt.ec3.at/services/management/updateXCipher" />
            <wsdl:input>
                <soap:body use="literal" />
            </wsdl:input>
            <wsdl:output>
                <soap:body use="literal" />
            </wsdl:output>
        </wsdl:operation>
        <wsdl:operation name="updateDocumentXCipher">
            <soap:operation
                soapAction="http://semcrypt.ec3.at/services/management/updateXCipherForRole" />
            <wsdl:input>
                <soap:body use="literal" />
            </wsdl:input>
            <wsdl:output>
                <soap:body use="literal" />
            </wsdl:output>
        </wsdl:operation>
        <wsdl:operation name="deleteXCipher">
            <soap:operation
                soapAction="http://semcrypt.ec3.at/services/management/deleteXCipher" />
            <wsdl:input>
                <soap:body use="literal" />
            </wsdl:input>
            <wsdl:output>
                <soap:body use="literal" />
            </wsdl:output>
        </wsdl:operation>
        <wsdl:operation name="deleteDocumentXCipher">
            <soap:operation
                soapAction="http://semcrypt.ec3.at/services/management/deleteXCipherForUser" />
            <wsdl:input>
                <soap:body use="literal" />
            </wsdl:input>
            <wsdl:output>
                <soap:body use="literal" />
            </wsdl:output>
        </wsdl:operation>
        <wsdl:operation name="selectXCipher">
            <soap:operation
                soapAction="http://semcrypt.ec3.at/services/management/selectXCipher" />
            <wsdl:input>
                <soap:body use="literal" />
            </wsdl:input>
            <wsdl:output>
                <soap:body use="literal" />
            </wsdl:output>
        </wsdl:operation>
        <wsdl:operation name="selectDocumentXCipher">
```

```
              <soap:operation
                 soapAction="http://semcrypt.ec3.at/services/management/selectXCipherForUser" />
              <wsdl:input>
                 <soap:body use="literal" />
550           </wsdl:input>
              <wsdl:output>
                 <soap:body use="literal" />
              </wsdl:output>
           </wsdl:operation>
555        <wsdl:operation name="insertUser">
              <soap:operation
                 soapAction="http://semcrypt.ec3.at/services/management/insertUser" />
              <wsdl:input>
                 <soap:body use="literal" />
560           </wsdl:input>
              <wsdl:output>
                 <soap:body use="literal" />
              </wsdl:output>
           </wsdl:operation>
565        <wsdl:operation name="updateUser">
              <soap:operation
                 soapAction="http://semcrypt.ec3.at/services/management/updateUser" />
              <wsdl:input>
                 <soap:body use="literal" />
570           </wsdl:input>
              <wsdl:output>
                 <soap:body use="literal" />
              </wsdl:output>
           </wsdl:operation>
575        <wsdl:operation name="deleteUser">
              <soap:operation
                 soapAction="http://semcrypt.ec3.at/services/management/deleteUser" />
              <wsdl:input>
                 <soap:body use="literal" />
580           </wsdl:input>
              <wsdl:output>
                 <soap:body use="literal" />
              </wsdl:output>
           </wsdl:operation>
585        <wsdl:operation name="selectUser">
              <soap:operation
                 soapAction="http://semcrypt.ec3.at/services/management/selectUser" />
              <wsdl:input>
                 <soap:body use="literal" />
590           </wsdl:input>
              <wsdl:output>
                 <soap:body use="literal" />
              </wsdl:output>
           </wsdl:operation>
595        <wsdl:operation name="insertRole">
              <soap:operation
                 soapAction="http://semcrypt.ec3.at/services/management/insertRole" />
              <wsdl:input>
                 <soap:body use="literal" />
600           </wsdl:input>
              <wsdl:output>
                 <soap:body use="literal" />
              </wsdl:output>
           </wsdl:operation>
605        <wsdl:operation name="updateRole">
              <soap:operation
                 soapAction="http://semcrypt.ec3.at/services/management/updateRole" />
              <wsdl:input>
                 <soap:body use="literal" />
610           </wsdl:input>
              <wsdl:output>
                 <soap:body use="literal" />
              </wsdl:output>
           </wsdl:operation>
615        <wsdl:operation name="deleteRole">
              <soap:operation
                 soapAction="http://semcrypt.ec3.at/services/management/deleteRole" />
              <wsdl:input>
                 <soap:body use="literal" />
620           </wsdl:input>
              <wsdl:output>
                 <soap:body use="literal" />
              </wsdl:output>
           </wsdl:operation>
625        <wsdl:operation name="selectRole">
              <soap:operation
                 soapAction="http://semcrypt.ec3.at/services/management/selectRole" />
              <wsdl:input>
```

```
                <soap:body use="literal" />
            </wsdl:input>
            <wsdl:output>
                <soap:body use="literal" />
            </wsdl:output>
        </wsdl:operation>
        <wsdl:operation name="insertKey">
            <soap:operation
                soapAction="http://semcrypt.ec3.at/services/management/insertKey" />
            <wsdl:input>
                <soap:body use="literal" />
            </wsdl:input>
            <wsdl:output>
                <soap:body use="literal" />
            </wsdl:output>
        </wsdl:operation>
        <wsdl:operation name="updateKey">
            <soap:operation
                soapAction="http://semcrypt.ec3.at/services/management/updateKey" />
            <wsdl:input>
                <soap:body use="literal" />
            </wsdl:input>
            <wsdl:output>
                <soap:body use="literal" />
            </wsdl:output>
        </wsdl:operation>
        <wsdl:operation name="deleteKey">
            <soap:operation
                soapAction="http://semcrypt.ec3.at/services/management/deleteKey" />
            <wsdl:input>
                <soap:body use="literal" />
            </wsdl:input>
            <wsdl:output>
                <soap:body use="literal" />
            </wsdl:output>
        </wsdl:operation>
        <wsdl:operation name="selectKey">
            <soap:operation
                soapAction="http://semcrypt.ec3.at/services/management/selectKey" />
            <wsdl:input>
                <soap:body use="literal" />
            </wsdl:input>
            <wsdl:output>
                <soap:body use="literal" />
            </wsdl:output>
        </wsdl:operation>
    </wsdl:binding>
    <wsdl:service name="ManagementService">
        <wsdl:port binding="tns:ManagementServiceSOAP"
            name="ManagementServiceSOAP">
            <soap:address
                location="http://semcrypt.ec3.at/services/management" />
        </wsdl:port>
    </wsdl:service>
</wsdl:definitions>
```

B.6 SecurityService.wsdl

The following listing represents the WSDL version of the SemCrypt security
service, which as such is part of the SemCrypt Security Service Framework
(SCSSF) [147]. Service operations are specifically tailored to the needs of
the SemCrypt Database Management System (SCDBMS) [93]. Chapter 6
presents an RMI as well as TCP interface to satisfy application requirements
that can not be met by XML Web service technology.

Listing B.6: SecurityService.wsdl

```
<?xml version="1.0" encoding="UTF-8"?>
<wsdl:definitions xmlns:soap="http://schemas.xmlsoap.org/wsdl/soap/"
    xmlns:tns="http://semcrypt.ec3.at/services/security"
    xmlns:wsdl="http://schemas.xmlsoap.org/wsdl/"
```

```
 5      xmlns:xsd="http://www.w3.org/2001/XMLSchema"
        name="SecurityService"
        targetNamespace="http://semcrypt.ec3.at/services/security">
        <wsdl:types>
            <xsd:schema
10              targetNamespace="http://semcrypt.ec3.at/services/security">
                <xsd:complexType name="AuthorizationException"></xsd:complexType>
                <xsd:complexType name="CipherException"></xsd:complexType>
                <xsd:complexType name="IndexClassifierType">
                    <xsd:sequence>
15                      <xsd:element name="index" type="xsd:string" />
                    </xsd:sequence>
                </xsd:complexType>
                <xsd:complexType name="DocumentClassifierType">
                    <xsd:sequence>
20                      <xsd:element name="documentId" type="xsd:string" />
                        <xsd:element name="pathValue" type="xsd:string" />
                    </xsd:sequence>
                </xsd:complexType>
                <xsd:complexType name="Classifier">
25                  <xsd:choice>
                        <xsd:element name="DocumentClassifier"
                            type="tns:DocumentClassifierType">
                        </xsd:element>
                        <xsd:element name="IndexClassifier"
30                          type="tns:IndexClassifierType">
                        </xsd:element>
                    </xsd:choice>
                </xsd:complexType>
            </xsd:schema>
35      </wsdl:types>
        <wsdl:message name="authorizeRequest">
            <wsdl:part name="pathValue" type="xsd:string" />
            <wsdl:part name="operation" type="xsd:string"></wsdl:part>
        </wsdl:message>
40      <wsdl:message name="authorizeResponse">
            <wsdl:part name="response" type="xsd:string" />
        </wsdl:message>
        <wsdl:message name="encryptValueRequest">
            <wsdl:part name="value" type="xsd:base64Binary"></wsdl:part>
45          <wsdl:part name="operation" type="xsd:string"></wsdl:part>
            <wsdl:part name="classifier" type="tns:Classifier"></wsdl:part>
        </wsdl:message>
        <wsdl:message name="encryptValueResponse">
            <wsdl:part name="encrypted" type="xsd:base64Binary"></wsdl:part>
50      </wsdl:message>
        <wsdl:message name="decryptValueRequest">
            <wsdl:part name="value" type="xsd:base64Binary"></wsdl:part>
            <wsdl:part name="operation" type="xsd:string"></wsdl:part>
            <wsdl:part name="classifier" type="tns:Classifier"></wsdl:part>
55      </wsdl:message>
        <wsdl:message name="decryptValueResponse">
            <wsdl:part name="decrypted" type="xsd:base64Binary"></wsdl:part>
        </wsdl:message>
        <wsdl:message name="encryptIdRequest">
60          <wsdl:part name="id" type="xsd:base64Binary"></wsdl:part>
            <wsdl:part name="operation" type="xsd:string"></wsdl:part>
            <wsdl:part name="classifier" type="tns:Classifier"></wsdl:part>
        </wsdl:message>
        <wsdl:message name="encryptIdResponse">
65          <wsdl:part name="encrypted" type="xsd:base64Binary"></wsdl:part>
        </wsdl:message>
        <wsdl:message name="decryptIdRequest">
            <wsdl:part name="id" type="xsd:base64Binary"></wsdl:part>
            <wsdl:part name="operation" type="xsd:string"></wsdl:part>
70          <wsdl:part name="classifier" type="tns:Classifier"></wsdl:part>
        </wsdl:message>
        <wsdl:message name="decryptIdResponse">
            <wsdl:part name="decrypted" type="xsd:base64Binary"></wsdl:part>
        </wsdl:message>
75      <wsdl:message name="SecurityServiceCipher_faultMsg">
            <wsdl:part name="fault" type="tns:CipherException"></wsdl:part>
        </wsdl:message>
        <wsdl:message name="SecurityServiceAuthorization_faultMsg">
            <wsdl:part name="fault" type="tns:AuthorizationException"></wsdl:part>
80      </wsdl:message>
        <wsdl:portType name="SecurityService">
            <wsdl:operation name="authorize">
                <wsdl:input message="tns:authorizeRequest" />
                <wsdl:output message="tns:authorizeResponse" />
85              <wsdl:fault name="fault"
                    message="tns:SecurityServiceAuthorization_faultMsg">
                </wsdl:fault>
```

```
          </wsdl:operation>
          <wsdl:operation name="encryptValue">
              <wsdl:input message="tns:encryptValueRequest"></wsdl:input>
              <wsdl:output message="tns:encryptValueResponse"></wsdl:output>
              <wsdl:fault name="fault"
                  message="tns:SecurityServiceCipher_faultMsg">
              </wsdl:fault>
          </wsdl:operation>
          <wsdl:operation name="decryptValue">
              <wsdl:input message="tns:decryptValueRequest"></wsdl:input>
              <wsdl:output message="tns:decryptValueResponse"></wsdl:output>
              <wsdl:fault name="fault"
                  message="tns:SecurityServiceCipher_faultMsg">
              </wsdl:fault>
          </wsdl:operation>
          <wsdl:operation name="encryptId">
              <wsdl:input message="tns:encryptIdRequest"></wsdl:input>
              <wsdl:output message="tns:encryptIdResponse"></wsdl:output>
              <wsdl:fault name="fault"
                  message="tns:SecurityServiceCipher_faultMsg">
              </wsdl:fault>
          </wsdl:operation>
          <wsdl:operation name="decryptId">
              <wsdl:input message="tns:decryptIdRequest"></wsdl:input>
              <wsdl:output message="tns:decryptIdResponse"></wsdl:output>
              <wsdl:fault name="fault"
                  message="tns:SecurityServiceCipher_faultMsg">
              </wsdl:fault>
          </wsdl:operation>
  </wsdl:portType>
  <wsdl:binding name="SecurityServiceSOAP"
      type="tns:SecurityService">
      <soap:binding style="document"
          transport="http://schemas.xmlsoap.org/soap/http" />
      <wsdl:operation name="authorize">
          <soap:operation
              soapAction="http://semcrypt.ec3.at/services/security/authorize" />
          <wsdl:input>
              <soap:body use="literal" />
          </wsdl:input>
          <wsdl:output>
              <soap:body use="literal" />
          </wsdl:output>
      </wsdl:operation>
      <wsdl:operation name="encryptValue">
          <soap:operation
              soapAction="http://semcrypt.ec3.at/services/security/encryptValue" />
          <wsdl:input>
              <soap:body use="literal" />
          </wsdl:input>
          <wsdl:output>
              <soap:body use="literal" />
          </wsdl:output>
      </wsdl:operation>
      <wsdl:operation name="decryptValue">
          <soap:operation
              soapAction="http://semcrypt.ec3.at/services/security/decryptValue" />
          <wsdl:input>
              <soap:body use="literal" />
          </wsdl:input>
          <wsdl:output>
              <soap:body use="literal" />
          </wsdl:output>
      </wsdl:operation>
      <wsdl:operation name="encryptId">
          <soap:operation
              soapAction="http://semcrypt.ec3.at/services/security/encryptId" />
          <wsdl:input>
              <soap:body use="literal" />
          </wsdl:input>
          <wsdl:output>
              <soap:body use="literal" />
          </wsdl:output>
      </wsdl:operation>
      <wsdl:operation name="decryptId">
          <soap:operation
              soapAction="http://semcrypt.ec3.at/services/security/decryptId" />
          <wsdl:input>
              <soap:body use="literal" />
          </wsdl:input>
          <wsdl:output>
              <soap:body use="literal" />
          </wsdl:output>
```

178   APPENDIX B. SERVICE INTERFACES

```
              </wsdl:operation>
          </wsdl:binding>
          <wsdl:service name="SecurityService">
              <wsdl:port binding="tns:SecurityServiceSOAP"
175               name="SecurityServiceSOAP">
                  <soap:address
                      location="http://semcrypt.ec3.at/services/security" />
              </wsdl:port>
          </wsdl:service>
180   </wsdl:definitions>
```

Bibliography

[1] C. Allen and L. Pilot. HR-XML: Enabling Pervasive HR e-Business.
http://www.gca.org/papers/xmleurope2001/papers/html/
s18-2b.html, May 2001.

[2] G. Alonso, F. Casati, H. Kuno, and V. Machiraju. *Web Services. Concepts, Architectures and Applications.* Springer-Verlag, Berlin, Heidelberg, 2004.

[3] S. Anderson, J. Bohren, T. Boubez, M. Chanliau, G. Della-Librera, B. Dixon, P. Garg, M. Gudgin, S. Hada, P. Hallam-Baker, M. Hondo, C. Kaler, H. Lockhart, R. Martherus, H. Maruyama, A. Nadalin, N. Nagaratnam, A. Nash, R. Philpott, D. Platt, H. Prafullchandra, M. Sahu, J. Shewchuk, D. Simon, D. Srinivas, E. Waingold, D. Waite, D. Walter, and R. Zolfonoon. Web Services Secure Conversation Language (WS-SecureConversation).
http://www-128.ibm.com/developerworks/library/
specification/ws-secon/, February 2005.

[4] S. Anderson, J. Bohren, T. Boubez, M. Chanliau, G. Della-Librera, B. Dixon, P. Garg, M. Gudgin, P. Hallam-Baker, M. Hondo, C. Kaler, H. L. R. Martherus, H. Maruyama, A. Nadalin, N. Nagaratnam, A. Nash, R. Philpott, D. Platt, H. Prafullchandra, M. Sahu, J. Shewchuk, D. Simon, D. Srinivas, E. Waingold, D. Waite, D. Walter, and R. Zolfonoon. Web Services Trust Language (WS-Trust).
http://www-128.ibm.com/developerworks/library/
specification/ws-trust/, February 2005.

[5] C. Anutariya, S. Chatvichienchai, M. Iwiahara, V. Wuwongse, and Y. Kambayashi. *Rules and Rule Markup Languages for the Semantic Web*, volume 2876, chapter A Rule-Based XML Access Control Model, pages 35–48. Springer-Verlag, Berlin, Heidelberg, 2003.

[6] Apache Software Foundation. Apache Tomcat.
http://tomcat.apache.org/, 2004.

[7] Apache Software Foundation. Welcome to JaxMe 2.
http://ws.apache.org/jaxme/, 2004.

[8] Apache Software Foundation. Apache Xindice.
http://xml.apache.org/xindice/, 2005.

[9] Apache Software Foundation. Web Services - Axis.
http://ws.apache.org/axis/, 2005.

[10] Apache Software Foundation. Apache WSS4J.
http://ws.apache.org/wss4j/, 2006.

[11] Apache Software Foundation. Welcome to XMLBeans.
http://xml.apache.org/xmlbeans/, 2006.

[12] Apache Software Foundation. Apache Cayenne. Object Relational
Mapping, Persistence and Caching for Java.
http://cayenne.apache.org/, 2007.

[13] Apache Software Foundation. Welcome to Apache Axis2/Java.
http://ws.apache.org/axis2/, 2007.

[14] Apache Software Foundation. Welcome to XML Security.
http://xml.apache.org/security/, 2007.

[15] A. Avižienis, J.-C. Laprie, B. Randell, and C. Landwehr. Ba-
sic Concepts and Taxonomy of Dependable and Secure Computing.
IEEE Transactions on Dependable and Secure Computing, 1(1):11–33,
January-March 2004.

[16] S. Bajaj, D. Box, D. Chappell, F. Curbera, G. Daniels, P. Hallam-
Baker, M. Hondo, C. Kaler, D. Langworthy, A. Nadalin, N. Nagaratnam, H. Prafullchandra, C. von Riegen, D. Roth, J. Schlimmer,
C. Sharp, J. Shewchuk, A. Vedamuthu, Ümit Yalçinalp, and D. Orchard. Web Services Policy Framework (WS-Policy). March 2006.
Version 1.2.
http://www-128.ibm.com/developerworks/library/
specification/ws-polfram/, March 2006.

[17] S. Bajaj, D. Box, D. Chappell, F. Curbera, G. Daniels, P. Hallam-
Baker, M. Hondo, C. Kaler, H. Maruyama, A. Nadalin, D. Orchard,

H. Prafullchandra, C. von Riegen, D. Roth, J. Schlimmer, C. Sharp, J. Shewchuk, A. Vedamuthu, and Ümit Yalçinalp. Web Services Policy Attachment (WS-PolicyAttachment). March 2006. Version 1.2. http://www-128.ibm.com/developerworks/library/ specification/ws-polatt/, March 2006.

[18] M. Bartel, J. Boyer, B. Fox, B. LaMacchia, and E. Simon. XML-Signature Syntax and Processing. W3C Recommendation 12 February 2002. http://www.w3.org/TR/xmldsig-core/, February 2002.

[19] K. Beck and E. Gamma. *Contributing to Eclipse. Principles, Patterns and Plug-Ins*. Addison-Wesley Professional, October 2003.

[20] A. Berglund, S. Boag, D. Chamberlin, M. F. Fernández, M. Kay, J. Robie, and J. Siméon. XML Path Language (XPath) 2.0. W3C Recommendation 23 January 2007. http://www.w3.org/TR/xpath20/, January 2007.

[21] T. Berners-Lee. Semantic Web - XML2000 - slide Architecture. http://www.w3.org/2000/Talks/1206-xml2k-tbl/slide10-0. html, 2000.

[22] E. Bertino, S. Castano, and E. Ferrari. Securing XML Documents with Author-X. *IEEE Internet Computing*, 5(3):21–31, 2001.

[23] E. Bertino, S. Castano, E. Ferrari, and M. Mesiti. Specifying and enforcing access control policies for XML document sources. *World Wide Web*, 3(3):139–151, 2000.

[24] E. Bertino and E. Ferrari. Secure and Selective Dissemination of XML Documents. *ACM Transactions on Information and System Security (TISSEC)*, 5(3):290–331, August 2002.

[25] S. Boag, D. Chamberlin, M. F. Fernández, D. Florescu, J. Robie, and J. Siméon. XQuery 1.0: An XML Query Language. W3C Recommendation 23 January 2007. http://www.w3.org/TR/xquery/, January 2007.

[26] D. Booth, H. Haas, F. McCabe, E. Newcomer, M. Champion, C. Ferris, and D. Orchard. Web Services Architecture. W3C Working Group Note 11 February 2004. http://www.w3.org/TR/ws-arch/, 2004.

[27] R. Bourret. XML and Databases.
http://www.rpbourret.com/xml/XMLAndDatabases.htm, 2005.

[28] D. Box, M. Hondo, C. Kaler, H. Maruyama, A. Nadalin, N. Nagaratnam, P. Patrick, C. von Riegen, and J. Shewchuk. Web Services Policy
Assertions Language (WS-PolicyAssertions). Version 1.0. December
18, 2002.
http://www-128.ibm.com/developerworks/library/
specification/ws-polas/, December 2002.

[29] J. Boyer. Canonical XML Version 1.0. W3C Recommendation 15 March
2001.
http://www.w3.org/TR/xml-c14n, March 2001.

[30] J. Boyer, D. E. Eastlake, and J. Reagle. Exclusive XML Canonicalization Version 1.0. W3C Recommendation 18 July 2002.
http://www.w3.org/TR/xml-exc-c14n/, July 2002.

[31] J. Boyer, M. Hughes, and J. Reagle. XML-Signature Xpath Filter 2.0.
W3C Recommendation 08 November 2002.
http://www.w3.org/TR/xmldsig-filter2/, November 2002.

[32] J. M. Boyer, D. Landwehr, R. Merrick, T. V. Raman, M. Dubinko, and
L. L. Klotz. XForms 1.0 (Second Edition). W3C Recommendation 14
March 2006.
http://www.w3.org/TR/xforms/, March 2006.

[33] T. Bray, J. Paoli, C. M. Sperberg-McQueen, E. Maler, and F. Yergeau.
Extensible Markup Language (XML) 1.0 (Fourth Edition). W3C Recommendation 16 August 2006, edited in place 29 September 2006.
http://www.w3.org/TR/REC-xml/, September 2006.

[34] R. Butek. Which style of WSDL should I use?
http://www-128.ibm.com/developerworks/webservices/library/
ws-whichwsdl/, 2005.

[35] L. M. Camarinha-Matos, editor. *Virtual Enterprises and Collaborative Networks*. Kluwer Academic Publishers, Toulouse, France, August
2004.

[36] J. Cao, L. Sun, and H. Wang. Towards Secure XML Document with Us-
age Contol. In Y. Zhang, editor, *7th Asia Pacific Web Conference (AP-
Web)*, pages 296–307, Shanghai, China, March 2005. Springer-Verlag
Berlin Heidelberg.

[37] H. L. Cardoso and E. Oliveira. Virtual Enterprise Normative Framework within Electronic Institutions. In *5th International Workshop on Engineering Societies in the Agents World (ESAW)*, Toulouse, France, October 2004.

[38] B. Carminati and E. Ferrari. AC-XML Documents: Improving the Performance of a Web Access Control Module. In *ACM Symposium on Access Control Models and Technologies (SACMAT)*, Stockholm, Sweden, June 2005. ACM.

[39] B. Carminati and E. Ferrari. Trusted Privacy Manager: A System for Privacy Enforcement on Outsourced Data. In *21st International Conference on Data Engineering Workshops (ICDEW)*, 2005.

[40] D. Chamberlin, D. Florescu, and J. Robie. XQuery Update Facility. W3C Working Draft 11 July 2006.
http://www.w3.org/TR/xqupdate/, 2006.

[41] S. Chatvichienchai, C. Anutariya, M. Iwaihara, V. Wuwongse, and Y. Kambayashi. Towards Integration of XML Document Access and Version Control. In F. Galindo, editor, *15th International Conference on Database and Expert Systems Applications (DEXA)*, pages 791–800, Zaragoza, Spain, September 2004. Springer-Verlag Berlin Heidelberg.

[42] A. B. Chaudhri, A. Rashid, and R. Zicari. *XML Data Management. Native XML and XML-Enabled Database Systems*. Addison-Wesley Professional, March 2003.

[43] Chiba Project. Chiba.
http://chiba.sourceforge.net/, February 2007.

[44] E. Christensen, F. Curbera, G. Meredith, and S. Weerawarana. Web Services Description Language (WSDL) 1.1. W3C Note 15 March 2001.
http://www.w3.org/TR/wsdl, March 2001.

[45] J. Clark. XSL Transformations (XSLT) Version 1.0. W3C Recommendation 16 November 1999.
http://www.w3.org/TR/xslt, November 1999.

[46] J. Clark and S. DeRose. XML Path Language (XPath) Version 1.0. W3C Recommendation 16 November 1999.
http://www.w3.org/TR/xpath, November 1999.

[47] codehaus. jaxen.
http://jaxen.org/, December 2006.

[48] CollabNet, Inc. GlassFish Community. Building an Open Source Java
EE 5 Application Server .
https://glassfish.dev.java.net/, 2006.

[49] CollabNet, Inc. javacc. Project Home.
https://javacc.dev.java.net/, 2006.

[50] CollabNet, Inc. xmldiff. Project home.
https://xmldiff.dev.java.net/, 2006.

[51] D. Connolly. Overview of SGML Resources.
http://www.w3.org/MarkUp/SGML/, March 2004.

[52] Cover Pages Hosted by OASIS. Project Management XML Schema
(PMXML).
http://xml.coverpages.org/projectManageSchema.html, March
2000.

[53] J. Crampton. Applying Hierarchical and Role-Based Access Control to
XML Douments. In *ACM Workshop on Secure Web Services*, Fairfax,
VA, USA, October 2004. ACM.

[54] E. Damiani, S. de Capitani di Vimercati, S. Paraboschi, and P. Samarati. A Fine-Grained Access Control System for XML Documents.
ACM Transactions on Information and System Security (TISSEC),
5(2):169–202, May 2002.

[55] E. Damiani, S. D. C. di Vimercati, S. Jajodia, S. Paraboschi, and
P. Samarati. Balancing Confidentiality and Efficiency in Untrusted
Relational DBMSs. In *10th ACM Conference on Computer and Com-
munications Security (CCE)*, Washington, DC, USA, October 2003.
ACM.

[56] E. Damiani, P. Samarati, S. D. C. di Vimercati, and S. Paraboschi.
Controlling Access to XML Documents. *IEEE Internet Computing*,
5(6):18–28, November-December 2001.

[57] G. Della-Libera, M. Gudgin, P. Hallam-Baker, M. Hondo,
H. Granqvist, C. Kaler, H. Maruyama, M. McIntosh, A. Nadalin,
N. Nagaratnam, R. Philpott, H. Prafullchandra, J. Shewchuk,
D. Walter, and R. Zolfonoon. Web Services Security Policy Language

(WS-SecurityPolicy). July 2005. Version 1.1.
`http://www-128.ibm.com/developerworks/library/`
`specification/ws-secpol/`, July 2005.

[58] S. DeRose, R. D. Jr., P. Grosso, E. Maler, J. Marsh, and N. Walsh. XML Pointer Language (XPointer). W3C Working Draft 16 August 2002.
`http://www.w3.org/TR/xptr/`, January 2002.

[59] S. DeRose, E. Maler, and D. Orchard. XML Linking Language (XLink) Version 1.0. W3C Recommendation 27 June 2001.
`http://www.w3.org/TR/xlink/`, June 2001.

[60] J. Dorn. Planning in virtual enterprises. *International journal of electronic business*, 2(5):557–565, 2004.

[61] J. Dorn. MOVE. Management and Optimization of business processes in Virtual Enterprises.
`http://move.ec3.at/`, 2005.

[62] J. Dorn and W. Schreiner. Security and Privacy Management in Virtual Enterprises. E-Commerce Competence Center (EC3), January 2007.

[63] S. Dustdar and W. Schreiner. A Survey on Web Services Composition. *International Journal of Web and Grid Services*, 1(1):1–30, January 2005.

[64] E. Dzakic. Tabellenkalkulation als verteilte Anwendung. Master's thesis, Institute of Software Technology and Interactive Systems, Electronic Commerce Group, Vienna University of Technology, 2007.

[65] D. Eastlake and J. Reagle. XML Signature WG.
`http://www.w3.org/Signature/`, July 2006.

[66] T. Eberhartl. Sichere Archivierung von SMS. Master's thesis, Institute of Software Technology and Interactive Systems, Electronic Commerce Group, Vienna University of Technology, 2007.

[67] Everware-CBDI Inc. Web Services Roadmap. Guiding the Transition to Web Services and SOA.
`http://roadmap.cbdiforum.com/reports/protocols/summary.php`, 2007.

[68] eXist. Open Source Native XML Database.
`http://exist.sourceforge.net/`, 2006.

[69] S. Farrell and S. Mysore. XML Key Management Working Group.
     http://www.w3.org/2001/XKMS/, 2004.

[70] S. Flesca, F. Furfaro, and E. Masciari. On the minimization of Xpath
     queries. In *29th International Conference on Vey Large Data Bases
     (VLDB)*, Berlin, Germany, September 2003.

[71] formsPlayer. formsPlayer.
     http://www.formsplayer.com/, 2007.

[72] I. Fundulaki and M. Marx. Specifying Access Control Policies for XML
     Documents with XPath. In *ACM symposium on Access control models
     and technologies (SACMAT)*, Yorktown Heights, New York, USA, June
     2004. ACM.

[73] E. Gamma, R. Helm, R. Johnson, and J. Vlissides. *Design Patterns.
     Elements of Reusable Object-Oriented Software.* Addison-Wesley Long-
     man, 1 edition, March 1995.

[74] S. K. Goel, C. Clifton, and A. Rosenthal. Derived Access Control
     Specification for XML. In *ACM Workshop on XML Security*, Fairfax,
     VA, USA, October 2003. ACM.

[75] G. Gottlob, C. Koch, and R. Pichler. Efficient Algorithms for Process-
     ing XPath Queries. In *28th International Conference on Very Large
     Data Bases (VLDB)*, Hong Kong, China, August 2002.

[76] G. Gottlob, C. Koch, and R. Pichler. XPath Processing in a Nutshell.
     In *ACM International Conference on Management of Data (SIGMOD)*,
     volume 32, pages 12 – 19, March 2003.

[77] K. Grün and M. Karlinger. Indexing Language. Technical report,
     Data and Knowledge Engineering, Johannes Kepler University of Linz,
     September 2006.

[78] K. Grün and M. Karlinger. Prototype specification. Technical report,
     Data and Knowledge Engineering, Johannes Kepler University of Linz,
     Linz, Austria, April 2006.

[79] H. Hacigümüş, B. Iyer, C. Li, and S. Mehrotra. Executing SQL over
     Encrypted Data in the Database-Service-Provider Model. In *ACM
     Special Interest Group on Management of Data (SIGMOD)*, Madison,
     Wisconsin, USA, June 2002. ACM.

[80] H. Hacigümüş, B. Iyer, and S. Mehrotra. Providing Database as a Service. In *18th International Conference on Data Engineering (ICDE)*, Washington, DC, USA, 2002. IEEE.

[81] S. Hada and M. Kudo. XML Access Contol Language: Provisional Authorization for XML Documents. http://www.trl.ibm.com/projects/xml/xacl/xacl-spec.html, October 2000.

[82] P. Hallam-Baker and S. H. Mysore. XML Key Management Specification (XKMS 2.0) Bindings. Version 2.0. W3C Recommendation 28 June 2005. http://www.w3.org/TR/xkms2-bindings/, June 2005.

[83] P. Hallam-Baker and S. H. Mysore. XML Key Management Specification (XKMS 2.0). Version 2.0. W3C Recommendation 28 June 2005. http://www.w3.org/TR/xkms2/, June 2005.

[84] B. C. Hammerschmidt, M. Kempa, and V. Linnemann. On the Intersection of XPath Expressions. In *9th International Database Engineering and Application Symposium (IDEAS)*, pages 49–57. IEEE, July 2005.

[85] F. Hirsch and M. Just. XML Key Management (XKMS 2.0) Requirements. W3C Note 05 May 2003. http://www.w3.org/TR/xkms2-req, May 2003.

[86] HR-XML Consortium. The independent platform for development of human resources XML vocabularies. http://www.hr-xml.org/, 2004.

[87] Human Genome Program of the U.S. Department of Energy Office of Science. Human Genome Project Information. http://www.ornl.gov/sci/techresources/Human_Genome/home.shtml, August 2006.

[88] IBM. Web Services Transactions specifications. http://www-128.ibm.com/developerworks/library/specification/ws-tx/, August 2005.

[89] T. Imamura, B. Dillaway, and E. Simon. XML Encryption Syntax and Processing. W3C Recommendation 10 December 2002. http://www.w3.org/TR/xmlenc-core/, December 2002.

[90] Internet 2. OpenSAML - an Open Source Security. Assertion Markup Language implementation. http://www.opensaml.org/, April 2007.

[91] D. Jordan and J. Evdemon. OASIS Web Services Business Process Execution Language (WSBPEL) TC. http://www.oasis-open.org/committees/tc_home.php?wg\ _abbrev=wsbpel, 2006.

[92] L. Kagal, S. Cost, T. Finin, and Y. Peng. A Framework for Distributed Trust Management. In *5th International Conference on Autonomous Agents*, Montréal, Canada, May 2001.

[93] M. Karlinger and K. Grün. Assembly of and Interaction between the SemCrypt System Components. Technical report, Data and Knowledge Engineering, Johannes Kepler University of Linz, July 2006.

[94] A. H. Karp. Authorization-Based Access Control for the Service Oriented Architecture. In *4th International Conference on Creating, Connecting and Collaborating through Computing*, Berkeley, CA, USA, January 2006. IEEE.

[95] A. H. Karp. Authorization-Based Access Control for the Services Oriented Architecture. Technical report, HP Laboratories, Palo Alto, January 2006.

[96] P. Kearney, J. Chapman, N. Edwards, M. Gifford, and L. He. An Overview of Web Services security. *BT Technology Journal*, 22(1):27–42, January 2004.

[97] A. Kemper and A. Eickler. *Datenbanksysteme. Eine Einführung*. Oldenburg, 2006.

[98] H. Koshutanski and F. Massacci. An Access Control Framework for Business Processes for Web Services. In *ACM Workshop on XML Security*, Fairfax, VA, USA, October 2003. ACM.

[99] J. K. Lee, S. J. Upadhyaya, H. R. Rao, and R. Sharman. Secure Konwledge Management and the Semantic Web. *Communications of the ACM*, 48(12):48–54, December 2005.

[100] C.-H. Lim, S. Park, and S. H. Son. Access Control of XML Documents Considering Update Operations. In *ACM Workshop on XML Security*, Fairfax, VA, USA, October 2003. ACM.

[101] L. Liu and S. Meder. Web Services Base Faults 1.2 (WS-BaseFaults). OASIS Standard, April 1 2006. http://docs.oasis-open.org/wsrf/wsrf-ws_base_faults-1.2-spec-os.pdf, April 2006.

[102] H. Lockhart, S. Andersen, J. B. nad Yakov Sverdlov, M. Hondo, H. Maruyama, A. Nadalin, N. Nagaratnam, T. Boubez, K. S. Morrison, C. Kaler, A. Nanda, D. Schmidt, D. Walters, H. Wilson, L. Burch, D. Earl, S. Baja, and H. Prafullchandra. Web Services Federation Language (WS-Federation). Version 1.1 December 2006. http://www-128.ibm.com/developerworks/library/specification/ws-fed/, July 2003.

[103] H. Lockhart, B. Parducci, and A. Anderson. OASIS eXtensible Access Control Markup Language (XACML) TC. http://www.oasis-open.org/committees/tc_home.php?wg_abbrev=xacml, 2003.

[104] M. Lorch, S. Proctor, R. Lepro, D. Kafura, and S. Shah. First Experiences Using XACML for Access Control in Distributed Systems. In *ACM Workshop on XML Security*, Fairfax, VA, USA, October 2003. ACM.

[105] B. Luo, D. Lee, and W.-C. L. andPeng Liu. QFilter: Fine-Grained Run-Time XML Access Control via NFA-based Query Rewriting. In *ACM Thirteenth Conference on Information and Knowledge Management (CIKM)*, Washington, DC, USA, November 2004. ACM.

[106] L. A. Maciaszek. *Requirements Analysis and System Design. Developing Information Systems with UML*. Addison-Wesley, 2001.

[107] K. D. Mann. *Java Server Faces in Action*. Manning Publications, 2005.

[108] D. L. McGuinness and F. van Harmelen. OWL Web Ontology Language Overview. W3C Recommendation 10 February 2004. http://www.w3.org/TR/owl-features/, February 2004.

[109] Microsoft Corporation. DCOM Technical Overview. http://msdn2.microsoft.com/en-us/library/ms809340.aspx, November 1996.

[110] Microsoft Corporation. Messaging Specifications Index Page. http://msdn2.microsoft.com/en-us/library/ms951268.aspx, 2007.

[111] Microsoft Corporation. XAML Overview.
`http://msdn2.microsoft.com/en-us/library/ms752059.aspx`,
2007.

[112] G. Miklau and D. Suciu. Containment and Equivalence for an XPath Fragment. In *21st ACM SIGMOD-SIGACT-SIGART Symposium on Principles of Database Systems (PODS)*, pages 65 – 76, Madison, Wisconsin, USA, June 2002.

[113] G. Miklau and D. Suciu. Controlling Access to Published Data Using Cryptography. In *29th International Conference on Very Large Data Bases (VLDB)*, Berlin, Germany, 2003.

[114] P. Mishra, H. Lockhart, E. Maler, P. Madsen, R. Philpott, S. Anderson, and J. Hodges. OASIS Security Services (SAML) TC.
`http://www.oasis-open.org/committees/tc\_home.php?wg\_abbrev=security`, 2004.

[115] P. Mitra, C.-C. Pan, P. Liu, and V. Atluri. Privacy-preserving Semantic Interoperation and Access Control of Heterogeneous Databases. In *ACM Symposium on InformAtion, Computer and Communications Security (ASIACCS)*, Taipei, Taiwan, March 2006. ACM.

[116] mozilla.org. XML User Interface Language (XUL).
`http://www.mozilla.org/projects/xul/`, 2007.

[117] M. Murata, A. Tozawa, and M. Kudo. XML Access Control Using Static Analysis. *ACM Transactions on Information and System Security (TISSEC)*, 9(3):292 – 324, August 2006.

[118] A. Nadalin. Web Services Security: Moving up the stack. New specifications improve the WS-Security model.
`http://www-128.ibm.com/developerworks/library/ws-secroad/`, December 2002.

[119] NetBeans. NetBeans IDE 5.5.
`http://www.netbeans.org/`, 2004.

[120] Network System Architects, Inc. GSM Security.
`http://www.gsm-security.net/`, 2006.

[121] OASIS. OASIS XML Common Biometric Format (XCBF) TC.
`http://www.oasis-open.org/committees/xcbf/`, August 2003.

[122] OASIS. OASIS Web Services Security (WSS) TC.
      http://www.oasis-open.org/committees/wss/, 2004.

[123] OASIS. UDDI.
      http://www.uddi.org/, 2006.

[124] Object Management Group (OMG). Catalog Of OMG CORBA/IIOP
      Specifications.
      http://www.omg.org/technology/documents/corba_spec\
      _catalog.htm.

[125] Object Management Group (OMG). Unified Modeling Language. UML
      Resource Page.
      http://www.uml.org/, 2007.

[126] Oracle. Oracle Berkeley DB Product Family. High Performance, Em-
      beddable Database Engines.
      http://www.oracle.com/database/berkeley-db.html, 2006.

[127] Oracle. Oracle TopLink.
      http://www.oracle.com/technology/products/ias/toplink/
      index.html, 2007.

[128] N. Qi and M. Kudo. XML Access Control with Policy Matching Tree.
      In S. D. C. di Vimercati, editor, *10th European Symposium On Research
      In Computer Security (ESORICS)*, pages 3–23, Milan, Italy, September
      2005. Springer-Verlag Berlin Heidelberg.

[129] L. Qin and V. Atluri. Concept-level Access Control for the Semantic
      Web. In *ACM Workshop on XML Security*, Fairfax, VA, USA, October
      2003. ACM.

[130] J. Reagle. XML Encryption WG.
      http://www.w3.org/Encryption/2001/, November 2005.

[131] Red Hat Middleware. Relational Persistence for Java and .NET.
      http://www.hibernate.org/, 2006.

[132] J. Rosenberg and D. Remy. *Securing Web Services with WS-Security.
      Demystifying WSSecurity, WS-Policy, SAML, XML Signature and
      XML Encryption.* Sams Publishing, 2004.

[133] B. Schneier. *Applied Cryptography. Protocols, Algorithms and Source-
      Code in C*, volume 2. John Wiley and Sons, January 1996.

[134] M. Schrefl, K. Grün, and J. Dorn. SemCrypt - Ensuring Privacy of Electronic Documents Through Semantic-Based Encrypted Query Processing. In *International Workshop on Privacy Data Management (PDM)*, Tokyo, Japan, April 2005. IEEE Computer Society.

[135] W. Schreiner. The Development of a Service Oriented System for Medical Image Conversion. Master's thesis, Institute of Scientific Computing, Group for Software Science, University of Vienna, November 2003.

[136] W. Schreiner. Integrating State-of-the-Art Web Services with a Medical Image Conversion System. Master's thesis, Institute of Scientific Computing, Group for Software Science, University of Vienna, November 2005.

[137] W. Schreiner. Stateful Web Services mit Apache WSRF. *entwickler magazin*, 4:136–140, June 2006.

[138] W. Schreiner. On Web Service Evolution Monitoring. In *3rd International Conference on Interoperability for Enterprise Software and Applications (I-ESA)*. Springer, March 2007.

[139] W. Schreiner. Security and Privacy Management in Service Oriented Architectures. In *3rd International Conference on Interoperability for Enterprise Software and Applications (I-ESA)*, Funchal (Madeira Island), Portugal, March 2007.

[140] W. Schreiner. SemCrypt Application Framework (SCAF). Technical report, E-Commerce Competence Center (EC3), 2007.

[141] W. Schreiner. SemCrypt Architectural Requirements. Technical report, E-Commerce Competence Center (EC3), 2007.

[142] W. Schreiner. SemCrypt Authentication and Authorization Framework (SCAAF). Technical report, E-Commerce Competence Center (EC3), 2007.

[143] W. Schreiner. SemCrypt Database Framework (SCDBF). Technical report, E-Commerce Competence Center (EC3), 2007.

[144] W. Schreiner. SemCrypt Encryption and Decryption Framework (SCEDF). Technical report, E-Commerce Competence Center (EC3), 2007.

[145] W. Schreiner. SemCrypt Native XML Database Integration. Technical report, E-Commerce Competence Center (EC3), 2007.

[146] W. Schreiner. SemCrypt Prototype Environment. Technical report, E-Commerce Competence Center (EC3), 2007.

[147] W. Schreiner. SemCrypt Security Services Frameword (SCSSF). Technical report, E-Commerce Competence Center (EC3), 2007.

[148] W. Schreiner. SemCrypt Web Services Framework (SCWSF). Technical report, E-Commerce Competence Center (EC3), 2007.

[149] W. Schreiner and S. Dustdar. Collaborative Web Service Technologies. In *Workshop on Challenges in Collaborative Engineering (CCE)*, Sopron, Hungary, April 2005. CCE.

[150] S. Schweigl. User-Interface Generator für SemCrypt Applikationen. Master's thesis, Institute of Software Technology and Interactive Systems, Electronic Commerce Group, Vienna University of Technology, 2007.

[151] T. Schwentick. XPath Query Containment. In *ACM International Conference on Management of Data (SIGMOD)*, volume 33, pages 101 – 109, Paris, France, March 2004.

[152] C. M. Sperberg-McQueen and H. Thompson. XML Schema. http://www.w3.org/XML/Schema, January 2007.

[153] R. Steele, W. Gardner, T. S. Dillon, and A. Erradi. XML-Based Declarative Access Control. In M. Bielikova, editor, *31st Annual Conference on Current Trends in Theory and Practice of Informatics (SOFSEM)*, pages 310–319, Liptovsky Jan, Slovak Republic, January 2005. Springer-Verlag Berlin Heidelberg.

[154] M. Strembeck and G. Neumann. An Integrated Approach to Engineer and Enforce Context Constraints in RBAC Environments. *ACM Transactions on Information and System Security (TISSEC)*, 7(3):392–427, August 2006.

[155] Sun Microsystems, Inc. Java Remote Method Invocation (RMI). http://java.sun.com/j2se/1.3/docs/guide/rmi/, 1999.

[156] Sun Microsystems, Inc. Java Authentication and Authorization (JAAS). http://java.sun.com/products/jaas/, 2007.

[157] Sun Microsystems, Inc. Java Cryptography Extension (JCE). http://java.sun.com/j2se/1.5.0/docs/guide/security/ CryptoSpec.html\#JceKeystore, 2007.

[158] Sun Microsystems, Inc. The Java ME Platform - the Most Ubiquitous Application Platform for Mobile Devices. http://java.sun.com/javame/index.jsp, 2007.

[159] A. S. Tanenbaum and M. van Steen. *Distributed Systems. Principles and Paradigms*. Prentice Hall, 2003.

[160] The Apache Software Foundation. The Apache Velocity Project. http://velocity.apache.org/, March 2007.

[161] The OpenTravel Alliance. OTA. http://www.opentravel.org/, 2004.

[162] The XML:DB Initiative. XUpdate - XML Update Language. http://xmldb-org.sourceforge.net/xupdate/, 2003.

[163] B. Thuraisingham. Directions for Security and Privacy for Semantic E-Business Applications. *Communications of the ACM*, 48(12):71–73, December 2005.

[164] W. Tolone, G.-J. Ahn, T. Pai, and S.-P. Hong. Access Control in Collaborative Systems. *ACM Computing Surveys*, 37(1):29–41, March 2005.

[165] UIML.org. Home of the User Interface Markup Language. http://www.uiml.org/, 2007.

[166] United Nations Economic Commission for Europe (UNECE). United Nations Directories for Electronic Data Interchange for Administration, Commerce and Transport. UN/EDIFACT Standard Directories. http://www.unece.org/trade/untdid/directories.htm, 2006.

[167] P. Vervest, E. van Heck, K. Preiss, and L.-F. Pau, editors. *Smart Business Networks*. Springer, November 2004.

[168] J.-Y. Vion-Dury and N. Layaida. Containment of XPath expressions: an inference and rewriting based approach. In *Extreme Markup Languages*, Montréal, Québec, Canada, August 2003.

[169] W3C. Latest SOAP versions. http://www.w3.org/TR/soap/, June 2003.

[170] W3C Architecture domain. Document Object Model (DOM).
http://www.w3.org/DOM/, January 2005.

[171] W3C Platform for Privacy Preferences Initiative. Platform for Privacy
Preferences (P3P) Project. Enabling Smarter Privacy Tools for the
Web.
http://www.w3.org/P3P/, October 2006.

[172] W3C Technology and Society domain. Semantic Web Activity. Resource Description Framework (RDF).
http://www.w3.org/RDF/, Jannuary 2007.

[173] W3Schools. DTD Tutorial.
http://www.w3schools.com/dtd/default.asp.

[174] J. Wang and S. L. Osborn. A Role-Based Approach to Access Control
for XML Databases. In *ACM Symposium on Access Control Models and
Technologies (SACMAT)*, Yorktown Heights, New York, USA, June
2004. ACM.

[175] Web Services Interoperability Organization. WS-I.
http://www.ws-i.org/, 2006.

[176] M. Weber. Anforderungsanalyse - Zsfg. Technical report, E-Commerce
Competence Center (EC3), March 2005.

[177] M. Weber. Overview on Scientific Objectives.
http://semcrypt.ec3.at/, June 2005.

[178] M. Weber. WP1 - Biotechnology. Technical report, E-Commerce Competence Center (EC3), April 2005.

[179] M. Weber. WP1 - eGovernment. Technical report, E-Commerce Competence Center (EC3), May 2005.

[180] M. Weber and P. Hrastnik. WP1 - HR-Szenario. Technical report,
E-Commerce Competence Center (EC3), March 2005.

[181] H. Werthner, M. Hepp, D. Fensel, and J. Dorn. Semantically-enabled
Service-oriented Architectures: A Catalyst for Smart Business Networks. In *Smart Business Networks Workshop*, Rotterdam, Nether-
lands, 2006.

[182] Wikipedia. Peer-to-peer.
http://en.wikipedia.org/wiki/P2P, 2006.

[183] XIML. eXtensible Interface Markup Language.
http://www.ximl.org/, 2007.

[184] XrML. The Digital Rights Language for Trusted Content and Services.
http://www.xrml.org/, 2005.

[185] I. Yagi, Y. Takata, and H. Seki. A Static Analysis Using Tree Automata for XML Access Control. In *3rd International Symposium on Automated Technology for Verification and Analysis (ATVA)*, volume 3707, pages 234–247. Springer LNCS, 2005.

[186] M. Yagüe. Semantic Access Control, A Semantics-based Access Control Model for Open and Distributed Environments.
http://www.lcc.uma.es/~yague/Semantics-basedAccessControl.html.

[187] M. I. Yagüe, M. del Mar Gallardo, and A. Manña. *Computer Security ESORICS 2005*, volume 3679, chapter Semantic Access Control Model: A Formal Specification, pages 24–43. Springer-Verlag, Berlin, Heidelberg, 2005.

[188] M. I. Yagüe and A. Maña. A Metadata-based Access Control Model for Web Services. *Internet Research*, 15(1):99–117, 2005.

[189] M. I. Yagüe, A. Maña, J. López, and J. M. Troya. Applying the Semantic Web Layers to Access Control. In *14th International Workshop on Database and Expert Systems Applications (DEXA)*, page 622, 2003.

[190] M. I. Yagüe and J. M. Troya. A Semantic Approach for Access Control in Web Services. In *EuroWeb 2002 Conference*, Oxford, UK, December 2002.

[191] Y. Zuo and B. Panda. Component Based Trust Management in the Context of a Virtual Organization. In *ACM Symposium on Applied Computing (SAC)*, Santa Fe, New Mexico, USA, March 2005. ACM.

Wissenschaftlicher Buchverlag bietet

kostenfreie

Publikation

von

wissenschaftlichen Arbeiten

Diplomarbeiten, Magisterarbeiten, Master und Bachelor Theses
sowie Dissertationen, Habilitationen und wissenschaftliche Monographien

Sie verfügen über eine wissenschaftliche Abschlußarbeit zu aktuellen oder zeitlosen Fragestellungen, die hohen inhaltlichen und formalen Ansprüchen genügt, und haben **Interesse an einer honorarvergüteten Publikation**?

Dann senden Sie bitte erste Informationen über Ihre Arbeit per Email an info@vdm-verlag.de. Unser Außenlektorat meldet sich umgehend bei Ihnen.

VDM Verlag Dr. Müller Aktiengesellschaft & Co. KG
Dudweiler Landstraße 125a
D - 66123 Saarbrücken

www.vdm-verlag.de